R. L Davy

Surgical Lectures

Delivered in the Theatre of the Westminster Hospital

R. L Davy

Surgical Lectures
Delivered in the Theatre of the Westminster Hospital

ISBN/EAN: 9783337021573

Printed in Europe, USA, Canada, Australia, Japan

Cover: Foto ©berggeist007 / pixelio.de

More available books at **www.hansebooks.com**

SURGICAL LECTURES.

SURGICAL LECTURES,

DELIVERED IN THE THEATRE OF THE

WESTMINSTER HOSPITAL,

BY

RICHARD DAVY, M.B., F.R.C.S.,

SURGEON TO THE HOSPITAL.

'Novelty is only in request; and it is as dangerous to be aged in any kind of course, as it is virtuous to be constant in any undertaking. There is scarce truth enough alive, to make societies secure; but security enough, to make fellowships accursed: much upon this riddle runs the wisdom of the world.'—DUKE VINCENTIO.

LONDON:

SMITH, ELDER, & CO., 15 WATERLOO PLACE.

1880.

THESE LECTURES

𝔄𝔯𝔢 𝔍𝔫𝔰𝔠𝔯𝔦𝔟𝔢𝔡

TO

MY FELLOW-WORKERS, COLLEAGUES AND STUDENTS,

OF THE

WESTMINSTER HOSPITAL;

WITH THE GOOD WISHES OF THEIR MUTUAL FRIEND,

RICHARD DAVY.

PREFACE.

I HAVE published the abstracts of my Lectures (many of which have already appeared in the Medical Journals) as a partial *résumé* of seven years' work as a Hospital Surgeon. The interest felt in the pursuit of original research is some compensation for the indifference of public discrimination. I trust that those who read these pages will not set me down as the touter of a '*big book*,' nor as the promoter of a narrow specialism.

33, WELBECK STREET,
    CAVENDISH SQUARE, LONDON, W.
      Midsummer Day, 1880.

CONTENTS.

a

SUBJECT V.

SUBJECT VI.

SUBJECT VII.

SUBJECT VIII.

SUBJECT IX.

SUBJECT X.

SUBJECT XI.

INDEX OF COMMUNICATIONS.

SURGICAL LECTURES.

SUBJECT I.

CLUB FOOT.

LECTURE I.

EXCISION OF THE CUBOID BONE FOR EXAGGERATED CASES OF TALIPES EQUINO-VARUS.

DELIVERED AT THE WESTMINSTER HOSPITAL, MARCH 28, 1876.

(*'British Medical Journal,' April* 29, 1876.)

GENTLEMEN,—I bring before your notice to-day the third case in which the cuboid bone has been removed for exaggerated and confirmed talipes equino-varus. The considerations that have led me to perform this operation are as follows:—1. The intractability and relapsing character of the deformity, notwithstanding division of tendons and plantar fasciæ; 2. The great trouble and expense to poor patients in procuring talipes instruments, and the necessity for prolonged surgical treatment; 3. The anatomical facts connected with the cuboid bone; and 4. The pathological evidence obtained from a case in which Mr. Barnard Holt had removed the cuboid bone for caries, unattended with club-foot. The two first-named considerations are matters of daily experience, and need not now occupy our time; but the anatomical facts are important, and worthy of statement.

The cuboid acts as a direct supporting block of bone to the two external toes (third and fourth metatarsal bones), having

B

for its base the anterior articular facet of the os calcis; it acts as a supporting wedge to the scaphoid and external cuneiform bones; and the ligaments attaching it to the os calcis are for the most part blended with other bones, notably the long calcaneo-cuboid, extending from the under tubercles of the os calcis to the metatarsal bases of the second, third, and fourth toes. There is but one muscle attached to it, viz., a tendinous slip of the flexor brevis pollicis. The tendon of the peroneus longus grooves its under surface, and supports it considerably as a sling.

The fourth consideration is a pathological one. Some few years ago, Mr. Barnard Holt removed nearly the whole of a left cuboid bone for caries, and last year the boy was in Matthew Ward for talipes valgus (the antagonistic deformity to varus); and I show you the cast taken from our museum. Now, on the same plan as you set a thief to catch a thief, so you artificially induce a talipes valgus to counteract varus by ablation of the cuboid.

Operation. — Use chloroform and Esmarch's bandage. Cut directly down on the cuboid from the outer side of the foot, through the indurated skin and bursa, and make the cut T-shaped by extending it over the dorsum of the foot; insert two stout wires, one into each flap, and use these as retractors during the operation. Having definitely exposed the upper and outer surfaces of the cuboid, screw the bone-forceps into the cancellous structure of the cuboid, and expand the blades until a firm leverage is gained. Then carefully divide the ligaments around the bone; wrench it out; and also avoid injuring the peroneus longus tendon beneath, by closely cutting on the bone itself with the knife. You will then notice the synovial membranes involved; and link your T-shaped aperture together by one or more sutures. No dressing whatever is to be applied to the wound,

FIG. 1.

but an internal foot-and-leg splint, with a gum and chalk bandage. Here are specimens of two intact cuboid bones that have been so excised.

The following extract I read to you from Adams's Jacksonian Prize Essay on Club-Foot, p. 251. 'Now that the anatomical conditions existing in congenital talipes varus, and the nature and extent of the structural changes and adaptations induced by the persistence of the deformity, are better understood than formerly, and we have at our command the improved instruments to the construction of which this knowledge has led, the operation of excision of the cuboid bone is not likely to be performed again in any case of talipes varus, even in adult life.'

In 1854, Dr. Little recommended this operation, and the late Mr. Solly, of St. Thomas's Hospital, performed it. The details of the operation are given ; the cuboid bone, and possibly portions of other bones, were also removed piecemeal by the gouge. A Scarpa's shoe was applied, and the result was ultimately less successful than had been anticipated by Mr. Solly.

Let us now consider the rationale of this treatment on a club-foot (talipes varus). The astragaloid division of the foot is untouched ; the metatarsal bones of the little and fourth toes are approximated to the os calcis ; so that the expanded tubercle on the fourth metatarsal rests on it. By fourth metatarsal I mean that of the little toe, because the great toe consists of three phalanges, and does not possess a metatarsal bone. The subsequent eversion and rotation of the foot anterior to the astragalus is much facilitated, because you concentrate your instrumental and manual energies on one hinge (astragalo-scaphoid), instead of two (astragalo-scaphoid, and calcaneo-cuboid) ; and, if you will pardon me for using a rough simile, it is far easier to wrench off the lid of a box that has support only from one hinge in the place of two. The block of cuboid, acting as a mechanical wedge, or key-stone, of the tarsal arch, is taken away ; the ligaments are more readily stretched or ruptured, while the cicatrising tissue intervening between metatarsal bones (third and fourth) and os calcis perpetuates approximation and contraction between this gap up to the point of bony contact. The following cases bear upon this question.

CASE I.—T. S., aged 15, was admitted on crutches, with exaggerated double talipes varus. The deformity was congenital, and he had been subjected to surgical discipline (off and on) since infancy. The cast of the deformity is shown prior to any operative procedure: the soles are completely approximated inwards, the dorsum of the feet looking outwards. On January 27, 1874, the left cuboid bone was excised, and is exhibited before you. The immediate result was striking; the sole of the foot could with force be placed in natural position; and on March 17, 1874, the left foot being in good figure, and wound healed, I cut out the right cuboid bone of the same patient. He did well, and left the hospital on May 27, 1874, a distinct plantigrade, wearing an ordinary boot, with special plantar hinge, and leg instruments to his waist. This boy, I am told, has joined a travelling show in the country, or you might have seen him promenading the purlieus of Westminster.

Case II.—On January 26, 1875, I excised both cuboid bones at one sitting, in a boy aged 9, for confirmed talipes equino-varus. The immediate result was not so marked as in my first case, but on strapping these two legs together with a leathern buckle, and so permitting the legs to act as an inside splint, each to each, during convalescence, the result was that both soles touched the ground. The boy left the hospital on May 26, 1875, walking very well with the aid of an instrument. I saw this boy on March 7, 1876; his instrument was worn out by incessant usage, his feet were slightly inverted, and he was recommended to reapply for admission, because the inner division of both plantar fasciæ was tense and cordlike.

Case III.—T. N., age 14, now occupies a bed in Luke Ward. On January 18, 1875, I excised this boy's right cuboid for most persistent talipes varus. You see how complete the reduction of the deformity is; how the sole can be placed in natural position; and assuredly this last case is the best of all the three. We have used a rectangular splint, the arm set off at right angles to the fibula, and sustaining a broad strap; this mechanism has exaggerated the leverage of a mimic peroneus longus muscle. The wound has healed satisfactorily without any dressings whatsoever.

On March 28, 1876, I removed a wedge-shaped piece from the tarsal arch; and acted upon the principle of obtaining bony union in place of fibrous. The boy is now under my care in the hospital, and bids fair to make a good recovery, with very marked diminution of deformity. Although I believe that all talipes cases have distinctly a pathological seat in the nervous system, I have purposely confined my remarks on this occasion to the aspect of its mechanical treatment for progressive ends.

Conclusions.—From this experience, limited as yet, we have good grounds for believing that excision of the cuboid bone is an operation to be admitted for talipes varus of a confirmed nature. The loss of this bone by no means prevents subsequent progression. The injury inflicted on the outer wall of the foot heals kindly. The subsequent treatment of a case by instrumentation is facilitated. I should not hesitate to remove the base of the outer metatarsal bone, in addition to the cuboid, if the severity of the deformity mechanically suggested such a course ; I do not advocate any intrusion on the os calcis. I am told that Mr. Davies-Colley, following a similar train of reasoning, at Guy's Hospital, has now recently operated successfully on a double club-foot. Admitting, therefore, that this treatment is still *sub judice,* I am glad to have the opportunity of ventilating these facts before a coming race of operating surgeons.

LECTURE II.

ABSTRACT OF A LECTURE ON CASES OF TALIPES.

DELIVERED MAY 18, 1877.

(*'British Medical Journal,'* December 15, 1877.)

GENTLEMEN,—During the winter session (1876-77), I have had the opportunity of showing the students typical cases of the four recognised forms of talipes, viz., varus, valgus, equinus, and calcaneus. The vulgar name for talipes, in any of its varieties, is club-foot; the surgical word is derived from talipedo, I walk on the ankles (talus=ankle, and pes=foot); varus =bow-legged ; valgus=knock-kneed ; equinus=belonging to horses ; calcaneus=belonging to the heel. Confusion constantly arises at the onset between the distortions *varus* and *valgus*. Practically, if you put your knees in the position of '*genu valgum*,' knock-kneed, your feet will as a consequence assume the distortion of *talipes valgus.* By attitudinising, you can readily impress the four cardinal points of the club-foot compass on memory. I have illustrated them on the diagram. (Fig. 2.)

Now as we suffer from a north-west or north-east wind, so do we also from talipes equino-varus or valgus: opposite extremes in club-foot do not meet: talipes varus and valgus are antithetical terms, and as far removed the one from the other as the east is from the west. The utmost aim of orthopœdic surgery is to render each case a plantigrade. The casts of each deformity are set out on the operating table. Club-foot depends

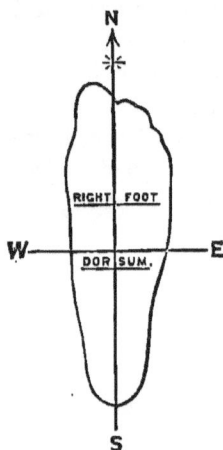

FIG. 2.

pathologically on a variety of causes, *e.g., mechanical,* from the result of burns, removal of bone; *mal-innervation,* both intra-uterine and extra-uterine; *malformation of parts; rachitis,* &c.; but the more I study these inaccuracies, the more I am convinced that the nervous system plays an important primary *rôle* in their origin and persistence; and that the muscular and tendinous structures are secondary exhibitors. Study, for example, this case.

M. F., aged 8, was admitted under my care on March 26, 1877, the subject of talipes varus of the left foot. On May 29, 1876, while he was running away, another boy in spite threw a glass bottle at his leg, which severed his peroneal nerve. The result of this accident was that gradually talipes varus has developed itself; flaccidity over the fibula; and inability to walk. Here is the scar; and we can eliminate the anterior tibial and cutaneous nerves, because the sensibility of the dorsum of his foot is unimpaired.

Let me next show you this boy; an exaggerated case of talipes varus of both feet, and congenital. This case is the third instance in which I have accurately removed a wedge-shaped portion of the tarsal arch for relapsed talipes varus. I have not yet received much support from the profession in performing this operation, Mr. Davies-Colley of Guy's Hospital being (so far as I know) the only surgeon who has as yet practised a somewhat similar operation with success. And here let me state once for all, that I am in no way indebted to any surgeon for the line of practice I have pursued, with the exception of the late Mr. Solly of St. Thomas' Hospital, who, in my opinion, was right in performing ablation of the cuboid, as a step in the right direction for the treatment of confirmed varus.

F. E., aged 12, Clapham, was admitted under my care on November 7, 1876. On November 14, 1876, I operated on the right foot, and he was convalescent on December 30. On January 16, 1877, I operated on his left foot, and he was convalescent on March 1, 1877. He shall to-day (May 18, 1877) walk, hop, run, and jump without any inconvenience and without any mechanical appliance, in the presence of you all. [After these evolutions had been gone through, the boy left the hospital.]

I will now state the line of argument that has led me to advise and practise this operation; and finish up by minutely describing the details of procedure. In a clinical lecture delivered in this theatre, March 1876, and printed in the 'British Medical Journal,' April 29, 1876, I gave my reasons for practising ablation of the cuboid, and illustrated practically its results in four cases. My dissatisfaction at the ordinary treatment of talipes varus by division of tendons and manipulation was based upon five years' experience at the Surgical Aid Society; where constantly relapsed cases were brought before my notice, which had been treated by our best orthopœdic surgeons.

In January, 1874, I commenced attacking the tarsal arch by excising the cuboid bone; and to-day I am ready to defend not only ablation of the cuboid, but an accurate removal of a wedge-shaped block of the tarsal arch. I have performed this operation three times in hospital practice with most excellent results; and I will now show you my original instruments, and the method of procedure. In the first place, you must fix most securely the foot on which you operate, and no dresser I have ever yet met with can hold a foot sufficiently steady for the precise use of the chisel. For this reason, I have introduced an ordinary portable vice into our operating theatre, with its jaws defended by the common cork clamps (as used by gunsmiths).

You must prepare yourselves in England to encounter opposition to any new project; and this vice in this theatre was

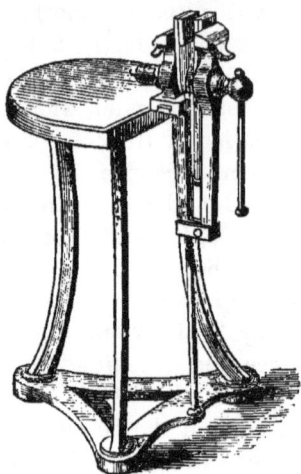

severely criticised before it was finally purchased. I have no hesitation in stating that a good vice for this operation is a *sine quâ non*. (Fig. 3.) Having put on Esmarch's bandage, accurately fix the leg and ankle in the vice; make an ⊥-shaped incision over the enlarged bursa overlying the cuboid; dissect back double door flaps; and insert stout silver wires to act as retractors: keep close to the bones, above and below, and clear a V-shaped space on the dorsum and sole of the foot, taking for the apex of the triangle the semilunar crease of skin that invariably exists on the inner side of the foot, the stereotyped line at

FIG. 3.

FIG. 4.

which inversion acts on the soft parts, as it were on a hinge:

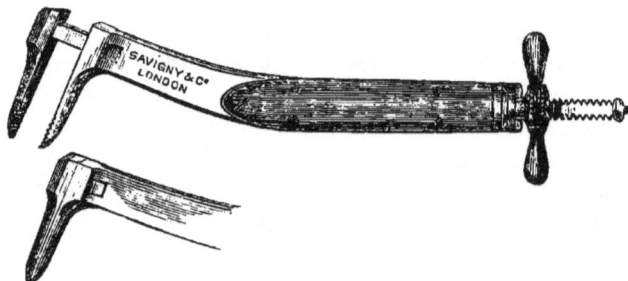

FIG. 5.

then use these chisels—painters' knives (fig. 4)—and accurately excise the wedge of the tarsus; this will embrace the cuboid, the head of the astragalus, part of the scaphoid, the base of the little metatarsal, and a chip of the external cuneiform bone ; use either of the bone-forceps (figs. 5 and 6) shown above for extracting the wedge. Approximate the gap, and chisel off right and left laminæ of bone until symmetry is restored ; rotation of the

FIG. 6.

phalangeal portion of the foot is also now performed, until the foot becomes plantigrade ; close the wound by tying the retracting wires together ; then fix the foot in this splint (figs. 7), and put up leg in a gum and chalk bandage over waterproof

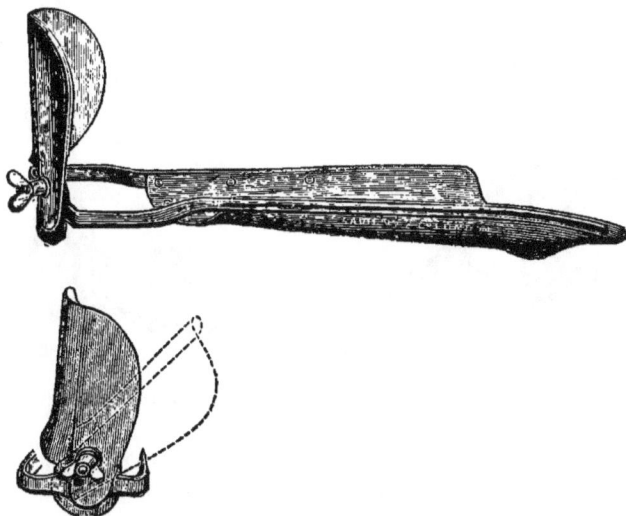

FIGS. 7.

splintage or flannel roller ; swing the foot so that the wound outside is dependent ; evert foot-piece until contour of foot is natural. The subsequent bleeding is not alarming ; the pain is by no means urgent ; swelling results ; synovial discharge follows ; and, so far as experience demonstrates, the wound is healed and the patient convalescent and able to stand in from six weeks to two months.

Now be good enough not to go away with the idea that every case of talipes varus is to be treated in this heroic fashion; this operation is a *dernier ressort* for obstinacy; an ordinary outside splint suffices for babies, with a gum and chalk bandage. In pedestrians, I have used the splint here engraved with admirable results (fig. 8). The boot with plantar hinge is in this instance applied to a valgus left foot; the crucial strap of elastic or leather inverts the foot; for varus the steel rod is to

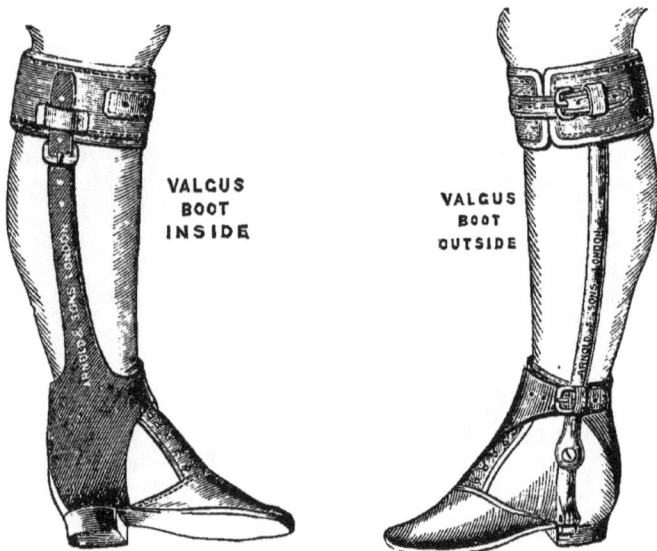

VALGUS BOOT INSIDE

VALGUS BOOT OUTSIDE

FIG. 8.

the inner side, and the crucial band to the outer; supplying an evertive force. No strap arrangement maintains discipline of natural contour more efficiently. This boot is by no means inelegant, and is made by Messrs. Arnold and Sons. The instruments are supplied by Savigny and Co.

The consideration of tenotomy, and the special treatment for talipes calcaneus and equinus, I will enter upon at a future date.

I will, lastly, read to you an extract from a letter received from the boy's father:

'I beg to take the liberty of offering you, in the name of

my wife and myself, sincere thanks for the great benefits my
son F. E. has received in the Westminster Hospital. All previous attempts to effect a cure having failed, we are delighted
to know that he is now able to run about with ease and comfort.'

[Since the delivery of this lecture, I have treated a case of
talipes equinus (nineteen years' duration) by an accurate excision
of the keystone of the tarsal arch.]

LECTURE III.

ON TALIPES EQUINUS AND CALCANEUS:

WITH CASES ILLUSTRATING NEW METHODS OF TREATMENT.

DELIVERED MAY 24, 1878.

('*British Medical Journal,*' *February* 15, 1879.)

GENTLEMEN,—Before entering upon our special subject this
day, I wish to draw your attention to two cases in Mark Ward,
of talipes varus, which have been most successfully operated on
in the way described by myself in the 'British Medical
Journal' of December 15, 1877 : cases in which common-
sense rules of cleanliness have been carried out, but without
any dressings, or any special antiseptic precautions. Both
were young lads; both had been most patient submitters to the
usual orthopædic practice; both had gradually got worse; and
both were unable to gain an honest livelihood by reason of the
deformity. The right foot was affected in each case. Their
recovery also was equally uniform. Each boy left his bed five
weeks after the operation; and both can now walk well without
apparent deformity.

You must remember that, in these cases of confirmed
talipes, the bones are structurally altered; and this fact
(amongst others) directed my attention to bone-operation, in
contradistinction to tendon or fascia-cutting. Take the case of
a deformed knee-joint with angularity of ten years' duration,
no amount of division of hamstrings will restore the limb to a
straight line; but an accurate removal of a wedge-shaped piece

of bone effects restoration immediately; or, supposing the
mast of a vessel to be permanently distorted, cutting away
portions of the rigging is useless; the mast itself must be subjected to the carpenter's skill before permanent straightness
and symmetry can be regained.

I here exhibit before you the facts which induced me to
resort to the removal of an accurate wedge from the tarsus.
On this dead man's foot I have opened freely from the outside
the transverse tarsal joint, and I have inserted a wedge of wood
into the gap. You now see how exactly the deformity of talipes
varus is produced, and how that nothing short of the ablation
of this wedge will restore symmetry. I am not aware that
this demonstration has ever been previously made; but it is an
excellent one, and makes the *rationale* of the operation at once
intelligible to any observer.

Talipes Equinus.—This deformity may be most simple
or severe; illustrative cases may be found from simple elevation of the heel, phalangeal progression, digital tip hobbling,
up to positive treading on the dorsum of the foot. In the
more severe cases, nothing short of amputation answers; and I
leave it to your discretion in individual application as to which
operation may be selected (Pirogoff, Syme, or Chopart). In
the simple variety, divisions of the tendo Achillis and plantaris
suffice.

The preparation and casts that I now show you are taken
from Gladhill's case of talipes equinus (nineteen years' duration),
in which I removed a wedge-shaped key-stone of the tarsal
arch. The man died, two weeks after the operation, from
septicæmia.

This cast (fig. 9) is a typical illustration of talipes equinus,
the heel itself not being so much at fault as the phalangeal end
of the foot; so I considered that, by dividing the tarsal arch
through its entirety, immediate restoration to flat-footedness
would result.

On November 20, 1877, I performed the operation, and
(conserving all the soft parts) excised, in the form of a wedge
(fig. 9, w.) from the dorsum to the sole, portions of the astragalus,
scaphoid, os calcis, and cuboid. The immediate result of the
operation is shown in this drawing, and by this cast (fig. 10).

This procedure not only mechanically restores symmetry,

but also so slackens the reins of the anterior group of tendons, that the toes can at once be put straight.

On examining the skeleton of this young man's foot, I am enabled to show you some interesting points in pathological anatomy. The arch of the foot is really very good ; the scaphoid abuts on the remnant of the astragalus, so that eventually bony union would have ensued, and the scaphoid thus would virtually have been absorbed into the astragalus; the os calcis would have united to the cuboid. The bones generally are small and light; the first phalanx (metatarsal bone) of the great toe is extremely puny ; while the metatarsal bone of the

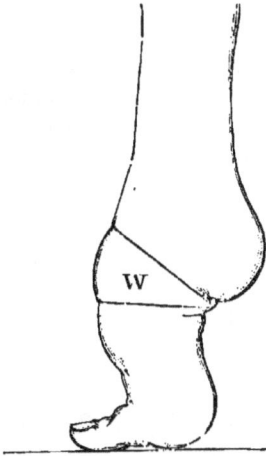 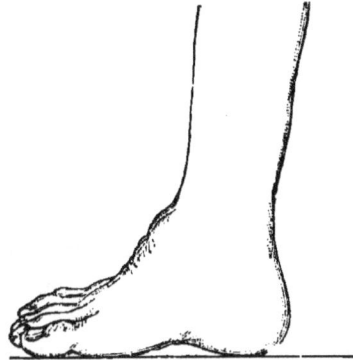

FIG. 9.   FIG. 10.

fifth rank is hypertrophied. On the dorsal and terminal end of each five metatarsal bones you may observe a completely new facette, on which rested the base of each phalanx ; the facette on the great toe is the least pronounced. The properly rounded tip of each metatarsal bone had no phalangeal base to be in contact with, and was never in use as a component part of these articulations. The bulk of the bony wedge removed consists of astragalus and os calcis, with thin slices off the posterior aspect of the cuboid and scaphoid bones.

This rectification of deformity by an operation applied to the transverse joint of the foot (*i.e.*, that immediately in front

of the os calcis and astragalus) supports strongly Mr. W. Adams's
view as to the spot at which the deformity hinges.

Before leaving the subject of talipes varus and equinus, I
will endeavour to sketch out a club-foot formula, which may
aid some of you in the consideration of these orthopœdic cases,
not only as affecting, on anatomical grounds, their causation,
but also as indications for treatment. In the same way the
triplet group of femoral muscles are arranged (viz., as extensors,
flexors, and adductors), so in the tibio-fibular group we have
to deal with extensors, flexors, and abductors. Of these, the
flexors are most powerful, then the extensors, and lastly the
peroneal section. Remember that the flexors of the leg or the
thigh are extensors of the foot and toes, and that the tendo
Achillis must be discussed separately; for its action is not only
to elevate the heel (the most movable point of its attachment),
but even to aggravate, as the case may be), the deformity of
talipes varus or valgus. And also bear in mind that the foot,
hinged at the ankle, has, by its own balance, a tendency to
droop forwards. We will now tabulate the varying muscles
and their nervous supply; and will indicate an excess of
nervous tonicity by the sign +, and a diminished nerve-
influence by the sign —. The principal muscles are: 1. The
sural group, ending in the tendo Achillis: 2. Tibialis posticus,
flexor longus digitorum, flexor longus pollicis, and the plantar
group; 3. Tibialis anticus, extensor longus digitorum, extensor
pollicis; 4. The peroneal group.

In taking each of these four groupings *seriatim*, we should
have as a sequence of our pathological conditions (*i.e.*, tonicity
or paralysis), in the first sural group, + equinus, — calcaneus;
in group No. 2 (flexors), + equinus, — calcaneus; in group
No. 3 (extensors), + calcaneus, — equinus; in group No. 4 (per-
onei), + valgus, — varus. In summarising these results,

+ 2 Eq.	+ Calc.	+ Valgus.
− 1 Eq.	− 2 Calc.	− Varus.

But, bearing in mind that the tibialis anticus and posticus both
act upon the tarsus anterior to the transverse tarsal joint, and
not on the toes, we must admit into the calculation

Tibialis anticus	+ Varus − Valgus.
Tibialis posticus	+ Varus − Valgus.

Then the club-foot formula stands as follows:—

	Equinus.	Varus.	Valgus.	Calcaneus.
+	2	2	1	1
−	1	1	2	2

Or, in plain English, the causation of equinus is opposed to calcaneus; that of varus to valgus; and the odds in favour of talipes equinus or varus being due to an excess of nerve-force are as two to one; while the odds in favour of calcaneus or valgus being due to paralysis is also as two to one. By tracing similarly the nervous supply to these muscles through the branches of the internal and external popliteal nerves, the same results, of course, follow. By combination, the chances of an equino-varus being due to nerve-tonicity is doubled; the chances of a calcaneo-valgus case being due to paralysis is also doubled, or, in mathematical expression, as four to two. I am well aware that this is taking one aspect alone of talipes formation; but I have found it to be a useful formula, and perchance you may find it the same.

Talipes Calcaneus.—I shall now say a few words on the treatment of talipes calcaneus, and on a mode of operating which I have acted upon twice in the wards of our hospital.

The primary object is to diminish tension on the extensor, and increase tonicity on the flexor group. This may be done by mechanisms, tenotomy, or by the galvanic current. But, having found in all severe cases of calcaneus the tendo Achillis presenting itself as a slack elongated band, I have thought it right to excise portions of this tendon; and, to make the argument practical, I will narrate and exhibit the following case.

P. S., aged 3, was admitted under my care into Chadwick Ward on October 29th, 1877, the subject of congenital talipes calcaneus of a marked character. On November 6, I excised one inch of his tendo Achillis by means of a straight cut parallel to the fibres of the tendon, and set up his foot in the state of extreme extension, so as to approximate the divided ends of the gastrocnemial group. The wound healed most readily; and, on December 4, 1877, I applied the actual cautery in four transverse lines, so as to induce cicatrisation and subsequent contraction. He left on December 30, 1877. On November 23, 1878, I was informed that the boy has recovered to the extent of discarding all mechanical support; but, owing to his mother having flitted, I was unable to obtain a personal interview.

I would, in conclusion, impress upon you that, since the introduction into surgical practice of these operations on the tarsal arch, no case of confirmed talipes varus exists that is not to be remedied by surgery; that talipes equinus is also equally under curative discipline; and that talipes valgus and calcaneus are conditions submitting readily to improvement by treatment which comprehends both general as well as strictly mechanical conduct.

ABSTRACT OF A PAPER

READ BY MR. DAVIES-COLLEY BEFORE THE MEDICO-CHIRURGICAL SOCIETY OF LONDON, ON OCTOBER 10, 1876; AND THE DISCUSSION THEREON.

A PAPER was read on a case of Resection of the Tarsal Bones for Congenital Talipes Equino-varus by J. N. C. Davies-Colley, Assistant-Surgeon to and Lecturer on Anatomy at Guy's Hospital. Notwithstanding the recent improvements in the treatment of club-foot, cases from time to time occur which, from the age to which the patient has attained, the rigidity of the tissues, and the altered shape of the bones, present insurmountable obstacles to a cure by the ordinary methods. It is not improbable that many of these cases might be successfully treated in the way which was adopted in the following instance: Edwin H——, aged twelve, was admitted into Guy's Hospital on May 8, 1875, under Mr. Cooper Forster. He was found to be the subject of severe talipes equino-varus, both feet being directed inwards, the soles backwards, the heels raised; and he could barely walk on account of the suppuration of bursæ which had formed on the back of the cuboid bones. He was kept in bed, and splints were applied. In September, 1875, Mr. Davies-Colley took charge of the case, and divided some tendons of the left foot. As very little advantage was thus gained, it was decided to take out a V-shaped piece from the tarsus. On the 12th of October this operation was performed. It was found

necessary to dissect out the cuboid bone, and then with knife
and saw to cut away portions of the os calcis, astragalus, scaphoid,
and cuneiform bones, together with the cartilage of the two
outer metatarsal bones. There was troublesome oozing of blood
after the operation, which was performed antiseptically; so a
sponge was put in, and the foot left in its old position for a
week. A peculiar splint was subsequently used in order to
twist the anterior half of the foot into its proper position. He
recovered rapidly, and on November 23 a similar operation was
performed on the right foot. Less than six weeks after the
second resection he was able to get about in a wheel chair. In
nine weeks he could walk with help. In ten weeks the wounds
were quite healed; all apparatus was left off, and he was able
to walk alone. He was shown at a meeting of the Medical and
Chirurgical Society in April last, and he could then run, jump,
and hop with considerable agility. He has since this time gone
to work, and in September a letter was received from his father,
which stated that he had walked six miles with but little fatigue.
The operation performed upon this boy was somewhat similar
to one which was suggested by Dr. Little, and has been
employed several times by Mr. Davy—viz., excision of the
cuboid bone. If, however, it be the object of the surgeon to
restore at once a severe case of talipes to the normal position,
it is necessary to cut away large portions of the adjacent bones,
as well as to remove the cuboid. This mode of procedure
appears to have been once adopted by Mr. Solly, but without a
very satisfactory result. The operation must be much simpler
when the foot is rendered bloodless by Esmarch's bandage.
The careful use of antiseptic precaution diminishes very much
the danger which would otherwise arise from the opening of so
many joints and synovial sheaths. Finally, the splint which
was afterwards used renders the subsequent moulding of the
foot into its proper relations a very easy matter. It is an
operation which may be adopted with great advantage where
other methods of treatment have failed, or when the patient
desires to avoid the long and painful treatment and the costly
apparatus which are required for the cure of the severe examples
of this deformity.

Mr. ADAMS, whilst congratulating the author upon the
favourable issue of this case, pointed out that, as a general law,

c

the operation of excision for cases of relapsed varus was totally unnecessary; since all such cases, no matter of how long standing, could be cured by mechanical means, aided by tenotomy, the tendo Achillis being the last tendon to be divided, on account of its contraction helping to remove the deformity. The argument that excision shortens time and avoids expensive apparatus was a doubtful one, as opposed to the risk incurred in opening joints and synovial cavities.

Mr. BRODHURST could not understand how, in a boy of 12, it was necessary to remove bone from the foot to remove a deformity which could be remedied even in the adult without such means. He remembered that in Mr. Solly's case some deformity remained twelve months after the operation.

Mr. RICHARD DAVY had seen so many patients treated by mechanical means, whose last state was worse than the first, that he had resolved to repeat Mr. Solly's operation (slightly modified)—namely, to excise the cuboid in cases of talipes. He had done this in three cases, and in the fourth had removed not only the cuboid, but a wedge-shaped piece of the tarsus.

Mr. CARR JACKSON thought it discreditable that so many cases of varus should be prevalent, and believed it to be due to the fact that the mechanical treatment was not continued for a long enough time. The treatment of a genuine case of talipes equino-varus lasts from infancy to adult life.

Mr. DAVIES-COLLEY, in reply, said that in his own account of his case Mr. Solly says that the operation was quite successful. There were a large number of successful cases among the poorer classes in whom, after operation, it was impossible to maintain the foot in position, and therefore relapses occurred. The dangers due to opening synovial cavities were diminished by antiseptic precautions. His operation was not simply excision of the cuboid, but of the whole tarsus.

CLINICAL SOCIETY OF LONDON.

FRIDAY, NOVEMBER 22, 1878.

MR. BRYANT exhibited a patient who had been the subject of talipes varus, and had been treated by the removal of a wedge of bone from the tarsus. This boy was 12 years of age. He had been under surgical treatment for the condition from infancy. When 5 years of age tenotomy had been performed, with some success; but as the Scarpa's shoe had caused pain, it was laid aside, and the deformity returned. On admission into Guy's Hospital the muscles of the leg were wasted, and the patient walked upon the outside of the foot, upon which had formed two large bursæ. Mr. Bryant removed a wedge-shaped piece of bone from the tarsus, as performed by Mr. Davies-Colley in October 1875. An incision was made across the dorsum of the foot, from a point corresponding to the tubercle of the scaphoid to the outer border of the cuboid; and a second incision along the outer border of the foot; the two incisions forming a ⌐ shape. The flaps were then turned back, and the tendons of the extensors divided. A spatula was introduced around the scaphoid bone, towards the sole of the foot, to protect the soft parts, and the lower section of the wedge of bone cut with a keyhole saw, one line of section extending across the dorsum of the foot from the scaphoid to the anterior border of the cuboid, the second bone section being made higher up; and a wedge with its apex corresponding to the scaphoid bone, and its base to the cuboid, one inch long, was thus cut away. After the operation the anterior half of the foot was readily brought round into position, and horse-hair drainage was employed. The temperature rose to 102 deg., but on the third day was down to 99·7 deg., with a pulse of 80. A small quantity of pus was evacuated by a puncture made into the skin, in a position corresponding to the apex of the wedge, and in other respects the wound healed rapidly. The boy exhibited presented a foot of good form, with a flat sole, on which he walked with comfort. The foot was somewhat shortened after the operation. The tendo Achillis had been cut, with the object of bringing down the heel, from which but little advantage had resulted. Mr. Bryant said that ablation of the cuboid had been suggested by Dr. Little in 1854, practised by Solly in 1854, and reintroduced to surgical notice by Mr. Richard Davy in 1874, upon which the operation now under consideration was a great improvement. He considered it also much better than Mr. Lund's operation, for the removal of the astragalus, which was performed in 1872, but which he thought might be useful where the equinus was worse than the varus.

Mr. RICHARD DAVY congratulated Mr. Bryant on the case of talipes varus that he had exhibited. Mr. Davy stated that he had operated in a similar way more frequently than any other surgeon. There was a class of confirmed and intractable cases of talipes that resisted all methods of treatment. In 1874, he re-opened the subject of the removal of the cuboid bone, which operation had been once performed by the late Mr. Solly in 1854. Thus for twenty years the operation had not only lain dormant, but had even been condemned by leading orthopœdic surgeons. During the years 1874 and 1875, he operated on five cases, and removed the cuboid; although the success resultant on this procedure was marked and striking, yet it was not quite perfect. He had proceeded on strictly experimental methods; and had not felt justified in interfering with the astragalo-scaphoid joint, until he had demonstrated that attacking the calcaneo-cuboid joint alone was insufficient. In April 1876, he published his experience; and an abstract of a clinical lecture was printed. In October 1876, Mr. Davies-Colley read his paper before the Medico-Chirurgical Society, which was the first intimation received at Westminster that Guy's Hospital had stolen a march on them. He had laid down the lines of operative procedure in a lecture published on December 15, 1877; and stated that the more closely those lines were followed, the better was the result gained. Mr. Davy showed the casts of his ninth patient; taken before and after the operation; and the result was very satisfactory. He considered that even now the operation was on its trial, and was opposed by many surgeons. Mr. Davy alluded to the criticism he had been subjected to in recommending these tarsal operations; but was content to abide by results.

Patients after the operation became absolutely plantigrades; the scar was small, and well out of the line of pressure; there was no possibility of relapse, and a symmetrical foot took the place of an unsightly and useless member.

ABSTRACT OF A PAPER

READ BEFORE THE MEDICAL SOCIETY OF LONDON, ON MARCH 11, 1878.

MR. RICHARD DAVY read a paper on ' Subdivision of the Tarsal Arch for Confirmed and Intractable Talipes.' He stated that his practice had been based on many experiments, the natural outcome of the operation performed by the late Mr. Solly in 1854. He had now the records of seven cases—viz., three in which the cuboid or both cuboids had been removed; three in which a wedge-shaped piece of the tarsal arch had been taken away for talipes equino-varus; and one in which a wedge had been excised for persistent talipes equinus. *Talipes Equino-varus*:—1. Boy, aged 15, January 27, 1874: left cuboid removed; recovery in seven weeks. March 1, 1874: right cuboid removed; recovery in nine weeks. 2. Boy, aged 14, January 18, 1875: right cuboid removed; recovery in ten weeks; result imperfect. 3. Boy, aged 9, January 26, 1875: both cuboids cut out at one sitting; recovery in three months. 4. Boy, aged 6, March 28, 1876: removal of wedge of tarsal arch; recovery in ten weeks. 5. Boy, aged 12, November 14, 1876: right tarsal arch divided; recovery in six weeks. January 16, 1877: left tarsal arch divided; recovery in six weeks. 6. Girl, aged 1 year 4 months, March 5, 1878: removal of portion of tarsal arch; progressing favourably. *Talipes Equinus :*—7. November 20, 1877: removal of wedge-shaped block of tarsus. December 5, 1877: death from septicæmia. Mr. Davy said his method of operating had been fully described, and insisted strongly that the operation should be limited to inveterate cases, which had not been amenable to milder treatment. Mr. Davies-Colley, Mr. Thomas Smith, and Professor John Wood, had each of them operated most successfully in such cases.

PAPER

READ BEFORE THE SURGICAL SECTION, BRITISH MEDICAL ASSOCIATION,
ANNUAL MEETING AT BATH, AUGUST 1878.

*Talipes Equino-Varus in an Adult treated successfully by the
Removal of the Astragalus, Scaphoid, and Cuboid Bones.*
By JAMES F. WEST, F.R.C.S. (Birmingham).

THE case was one of aggravated talipes equino-varus in a
woman aged 23, on whom Mr. West had operated by the plan
suggested by Mr. Richard Davy of the Westminster Hospital.
Models of the limb taken before and subsequently to the
operation, and also the tarsal bones, the cuboid, astragalus, and
scaphoid, which had been removed at the time of the operation
(May 19, 1878), were shown. The result, as seen ten weeks
after the operation, was eminently successful. Mr. West considered that, although tenotomy was the most suitable operation
in infants and young children, in adults, or in cases where
tenotomy had been tried and failed, resection of a wedge-
shaped portion of the tarsal bones was an excellent operation;
and that, in future, no case of talipes, whatever might be the
age of the patient or the severity of the deformity, need be
looked upon as hopeless. The use of Esmarch's bandage, and
the antiseptic method of subsequent treatment, in operative
procedures of this kind, deprived them of the risks which such
an interference with the tarsus and its numerous articulations
would otherwise involve.

SUBJECT II.

SPINAL CURVATURE.

LECTURE I.

OBSERVATIONS ON THE TREATMENT OF SPINAL CURVATURE.

('*Practitioner,*' *March* 1872.)

GENTLEMEN,—The treatment of spinal curvature may be essentially subdivided into—firstly general, and secondly mechanical, treatment.

Under the first count are included rest, sea-side air, strengthening food, oleum morrhuæ, careful nursing, and such like ; and the late Sir Benjamin Brodie concentrates the essence of such treatment in advising a couch pleasantly situate near the sea-beach ; indeed, it is useless to undertake the treatment of this deformity without paying marked deference to the general means of cure ; but especial prominence has been given in this short paper to the local and mechanical means of surgical assistance.

Under the second count are included local and mechanical means.

The utmost importance must be conceded to the recumbent posture at an early stage of the deformity. This necessity for early rest is in many instances entirely overlooked by the parents ; many a child is unnecessarily tormented by an anxious mother, who runs from one Orthopœdic Institution to another, in the mistaken delusion of thus obtaining the best treatment. The poor child protests, and shows its sufferings by its peevishness and groans ; the mother contents herself with the empty self-congratulation of having exhibited her offspring to goodness knows how many doctors. Nothing can so effectually give rest and ease to the diseased vertebral column as this apparently simple recommendation of the recumbent posture; but in reality, amongst the rich this treatment represents a couple of extra servants, amongst the poor it involves an impossibility.

Children again, not being aware of the importance of rest, are with difficulty kept lying down; probably the easiest means of insisting on this principle of rest with them is to net over their cribs; whilst among the hunchbacks at or about the time of puberty there exists such a refractory impatience of restraint and such precocity as to baffle the good intentions of any surgeon. Before leaving the subject of the recumbent posture, let especial stress be laid upon its importance in cases of cervical curvature; so as to avert any secondary implication of either the medulla oblongata, or roots of the phrenic nerves.

The difficulty then of restraining these cases of spinal curvature led to the employment of spinal instruments, on the principle of relieving the vertebral column of superincumbent weight, while freedom was allowed for taking sufficient exercise to maintain a certain degree of vigour.

What can the surgeon fairly expect from the use of a spinal instrument? Simply support, and a correction of the tendency to increased deformity. As a gardener supports the delicate stem of a plant by a firm stake; or as in young fir plantations side support and an upward direction and shelter are ensured by adjoining stems; so the surgeon uses a spinal instrument to shelter, support, and as it were coax the feeble spine into its healthy, natural position.

Let me now express my strong dissent to the too universal application of spinal instruments. Surgeons weekly receive application for spinal instruments, where no spinal disease exists, and where the appliance (if granted) would but tend to increase the deformity.

Let me further object to and expose a rather numerous class of individuals, who foolishly believe that their duty to their deformed charge has been performed as soon as the victim has been encased in a spinal instrument, and thus seek to shift the onus of treatment from their own to the surgeon's shoulders.

Still further, objection must be taken to the intrusive desire of any instrument-maker to complicate the essential simplicity of a spinal instrument: as a rule, the more movements, the more pay for the instrument-maker; but the more movements, the less relief for the patient.

Complicated movements, if kept in action, must guarantee much interference; if unused, such movements are quite unnecessary.

Side plates are certainly advantageous, if manual support to the projecting ribs and transverse processes gives relief to the patient; and of all elevating principles that have been applied to the crutch of spinal instruments there is none so easy and so practically useful as the principle used by Sayre for extension in cases of morbus coxæ. It allows elevation or depression to be performed easily, safely, and advantageously, either by the surgeon or patient; many a sore axilla will be saved, and much more support (consecutive on the growth or improved condition of the vertebral column) will be gained by the further use of this elegant mechanism.

The natural cure of these deformities consists in bony anchylosis of the bodies and laminæ of the vertebræ; and the frequency of bony nodules being found on bodies of the vertebræ demonstrates how ready nature is to throw out support for a feeble spine: yet even in old permanent fixtures of angular curvature of the spine you may still see movements on the instrument worn; *i.e.*, the officious surgeon endeavouring to undo what nature has wisely done. Let me once more insist therefore on the strict simplicity of a spinal instrument, as an agent of support as opposed to coercion.

To summarise this sketch. Grant attention to the sterling value of an early correct diagnosis; good general treatment; the importance of rest; the recumbent posture, and mechanism only as supportive agents.

1. For recent cases with advancing deformity, general treatment, rest, recumbent posture: as nature regains strength, and the bony deposit is being organised, mechanical support, and the sparing adjustment of spinal movements.

2. In chronic cases with stationary deformity, general treatment and mechanical support.

3. In hysterical cases, chloroform must be administered; moral control and physical exercise employed; and a full exposure given to any smack of deception.

4. In weakly constitutions with slight deformity, tonic treatment, sea-side baths, and correction of faulty tendencies.

LECTURE II.

CLINICAL LECTURE ON THE TREATMENT OF SPINAL CURVATURE BY HAMMOCK SUSPENSION, AND THE APPLICATION OF THE PLASTER, FELT, OR GLUE JACKET.

DELIVERED MAY 18, 1880.

('British Medical Journal,' June 26, 1880.)

GENTLEMEN,—In general practice, no cases have given the practical surgeon greater trouble in treatment than those of spinal curvature. He has to combine the general treatment of such cases with instrumentation; and, until the publication of Sayre's work, this combination was most inefficiently performed. Poverty of the patient's friends, intractability of the patient, importunity of the surgical mechanician, and a rapidly progressive disease, each drove the surgeon to his wit's end; and I say it is only since Sayre lifted the surgeon out of the grip of the machine-maker, and made him comparatively independent of the patient's surroundings, that real progress can be reported. The secret of Sayre's success is so well known, and his plan has been so publicly demonstrated by writings and on actual cases, that it is superfluous here to explain his mechanisms; but I advise those of you who have not yet read his book to carefully peruse it. 'On Spinal Disease and Spinal Curvature: their Treatment by Suspension and the use of Plaster-of-Paris Bandage' (Smith, Elder, and Co., 15, Waterloo Place, London. 1877). The work is written in Sayre's graphic style, and reflects much credit upon one of the most ingenious American surgeons. But, guided by my own experience, I must take exception to one or more points advocated by Sayre, and these exceptions will apply principally—1. To the risk and personal discomfort, both to the patient and surgeon, of the tripod suspension; 2. To the cracking, creasing, and consequent insufficiency of the support, due to the necessary shift of the patient's surroundings; 3. To the weight of the plaster-of-Paris. With regard to the risk of suspension by the chin and arm-

pits, I have no wish to play the part of an alarmist, nor to unduly exaggerate the dangers of extension to the spine; but, as
a pupil of the late John Hilton, I ask if any surgeon who
values past surgical records can carefully read Lecture V. in
his work on ' Rest and Pain,' and be ready to fearlessly suspend
cases of cervical caries ?

Noticing, moreover, how Nature expands into such massive
bony shields the laminæ of diseased vertebræ, and unites them,
on the principle of *vis unita fortior*, I fail to see how even
carefully regulated hanging can assist her in the process; and
my answer to those surgeons who state that they stop suspension the moment that pain is felt, is, 'Your interference
has arrived one moment too late.' As to the principle of
treating these spinal curvatures, I have always illustrated it
in my surgical lectures by placing a flexible india-rubber column
within a glass cylinder. The flexible tube represents the body
and diseased spine; the glass cylinder the rigid columnar support or jacket; and I believe that the more you allow these
diseased structures to remain in the recumbent posture, the
more closely do you approach the natural pose of rest, or, at any
rate, you bid fair to accomplish the maximum of good with
the minimum chance of evil.

In 1876, the year of the International Medical Congress at
Philadelphia, Professor Sayre did me the honour, in New York,
of inviting me to visit his office and showing me in detail his
plan on a case of angular curvature. I was at the time impressed with the disagreeableness of the suspension, and mistrusted its absolute necessity. On resuming orthopœdic practice
in London, I completed a series of experiments by suspending
patients in hammocks; and I venture to express my opinion
that, in future, more spinal curvatures will be fortified in the
manner I am now about to describe than after the suspension-
method of Sayre.

A piece of strong canvas is procured, longer than the patient's
height; and the arms are passed through two slits in the canvas at suitable points, so that, in the first instance, a loose
canvas long apron, with ends one turned downwards over the
chest and the other on the floor, fits around the front and sides of
the body. This apron is then removed from the patient, and a
vest applied, of thicker material and far more open mesh than

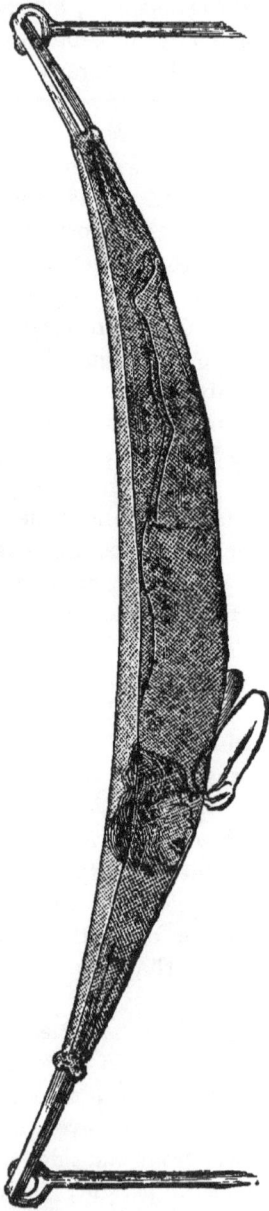

FIG. 11.

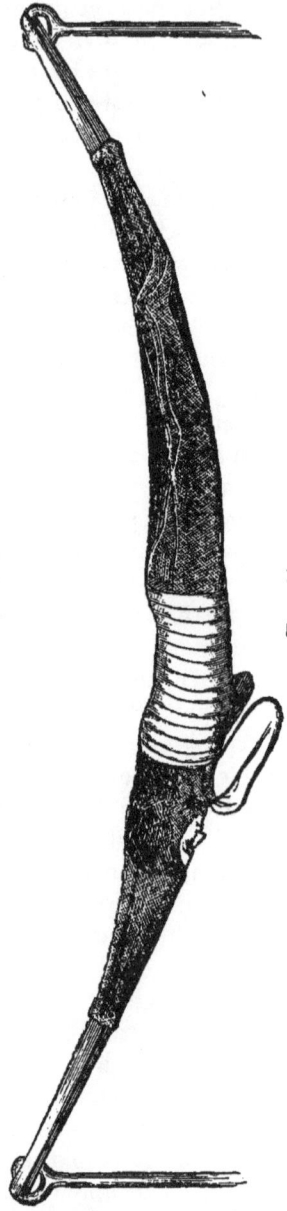

FIG. 12.

those usually supplied by the surgical instrument makers. The canvas hammock is next slung, at two fixed points, by attaching its two folded ends with two stout bandages: and the surgeon should test its bearing power by the weight of his own body. The patient is placed in the canvas as represented in this engraving. No dinner-pad is required, because the manipulations are performed after a good meal. (Fig. 11.)

At this stage, an aperture in the hammock is made over the patient's lips, to permit free breathing and conversation. The patient is then finally localised in position, according to the variety of spinal curvature (extension by the head, arms, and legs being applied by those surgeons who deem it to be necessary), and the surgeon leisurely applies the plaster-of-Paris or other fixing material, including the canvas, which, of course, has been accurately cut to shape the dorsal contour. After the bandages have been carefully laid on (and the ease of so doing is very great in this position), the patient presents the following appearances. (Fig. 12.)

The free current of air around the patient's body, and, if the surgeon please, the hammock's suspen-

FIG. 13.—Patient in the erect posture ; the upper and lower ends of the canvas having been neatly trimmed off.

sion near to a fire, facilitates the regular and simultaneously complete drying of the plaster; and so very comfortable are young children in these hammocks, that they either enjoy the swinging motion, or not unfrequently fall asleep. When the bandage has firmly set (and not before, for the patient can remain swung for any reasonable space of time), the whole hammock and patient are taken down; the superfluous ends are neatly cut off with scissors, and the general effect is shown in the sketch taken from my last patient (fig. 13); so that in these instances it may be literally said, and without irreverence, they take up their bed and walk; the canvas remnant acting as an accessory vest to the patient's frame.

I am ready to endorse much that has been written by Dr. Walker, Surgeon to the Peterborough Infirmary, on the utility of the recumbent posture ('British Medical Journal,' March 1, 1879); but think that my method negatives No. 4 conclusion in his summary: 'The only way in which such a jacket can be applied with the patient recumbent is by the method which I have demonstrated.'

Mr. Cocking, the well-known felt manufacturer for spinal jackets, attended one of my demonstrations in the theatre of Westminster Hospital on May 18, 1880; he considers that this hammock-suspension may prove very convenient for the application of his felt supports; as yet he has only tried it in one instance. This hammock-suspension, then, in my own practice obviates all the inconveniences and risk to patients; secures the advantages of Sayre's method; facilitates the manual procedures of the surgeon; permits of quick and equable drying; and necessitates no change of posture between the application of the supporting bandage and its final setting.

The third point is the weight of the plaster-of-Paris. I am experimenting now on felt bandages saturated with a quickly drying solution, but will defer judgment on them until my experience is confirmed by further practical trial.

I can highly recommend a solution of plaster-of-Paris and hair intermixed. You have seen me put up many cases of talipes with the mixture; the hair plays an important intertangling part, and from its elasticity and lightness, a very easy and accurately-moulded cast results. This species of felt packing will undoubtedly supplant the plaster-of-Paris, on the sole argument of its specific weight. Nothing can go against the grain of a thoughtful surgeon more than to handicap an already puny patient with a heavy load, for most of these clients with paralysed muscles can scarcely carry their own weight, much less that of even a few ounces extra.

One word, in conclusion, on the subject of portability and expense. Any country surgeon can carry hammock and bandages in his coat pocket, and so operate easily at the child's home instead of its being brought to the tripod. The cost of the hammock is under one shilling, and it may be extemporised out of a common sheet or a long night-shirt. I can truthfully state that, although we possess the complete tripod and apparatus at the Westminster Hospital, and have used it seldom, its joints are become very rickety, and it may yet become more impaired by desuetude, giving way to the simpler, but equally practical, system of hammock-suspension.

SAYRE'S PLASTIC JACKET.

COMMUNICATION FROM L. B. MASON, ESQ., OF PONTYPOOL, L.R.C.P.,
TO THE 'BRITISH MEDICAL JOURNAL,' JULY 31, 1880.

I HAVE a little patient, six years old, who, in consequence of Pott's spinal disease and psoas abscess, requires a jacket renewed about every three or four months. Hitherto, I have succeeded in the application by suspending from the armpits. This has been a tedious and unsatisfactory state of things, as I had no suspension apparatus, and had to rely upon an adult holding him up. After reading Mr. Davy's lecture in the *Journal*, I determined to try his plan; and, by means of two staples driven into the wall on opposite sides of the room, and a few yards of unbleached calico, I succeeded in applying a jacket which simply exceeded my anticipations—did not fatigue either patient or operator, and, above all, placed the little sufferer in a position as upright as a dart. After the application, I left him swinging about for his amusement for about two hours, and by that time the jacket was firm.

SUBJECT III.

INVALID HAMMOCKS.

—————

ABSTRACT OF A CLINICAL LECTURE ON THE USE OF HAMMOCKS IN SURGICAL PRACTICE.

DELIVERED JUNE 12, 1875.

(*'British Medical Journal,'* June 26, 1875.)

GENTLEMEN,—I introduce to your notice to-day Seydel's hammock, and wish to tell you the line of thought that has led me to employ it in surgical practice, and then to direct your attention to its more extended application. Two of these hammocks (a child's and an adult's) are now swung in this theatre ; and the illustrated sheet gives you a general idea of their usefulness, as well as a description of their material, size, weight, price, &c. At the Surgical Aid Society, I have for the last year used hammocks for the treatment of spinal curvature, because it seemed to be a most excellent method of giving rest to the spinal column, and guarantees an immediate enforcement of hygienic conditions. Bad landlordism and good surgery are sworn enemies ; how often are surgeons called to see a pitiable humpback crying in a dark corner of a squalid room, or sitting up, hands on knees, in some miserable tenement ! Do not be misled by supposing that a steel machine can put this in order, but direct your attention particularly towards maintaining the spinal column at rest, adopt measures calculated to increase the patient's vigour, and pay not too much attention to the actual deformity. Now, think for a moment how Nature grants immunity from accident and rest to spinal cord. Anatomy teaches us that the spinal cord affords an example of a suspended system of nerves resting on a fluid contained in a swung fibrous tube. Place, therefore, your patient on a water or air-cushion in one of these hammocks, and sling your hammock as near the window as convenient; then whatever little fresh air, light, sunshine, and rest can be gained in a rookery will be utilised; and, should an instrument of support be considered supplementary, further deformity is prevented during the time for carrying out mechanical detail. The patient's friends constantly pester surgeons with remarks on the deformity. Truly it is an index to the extent of bone dilapidation; but it need not occupy your thought any more than telegraphic engineers repairing a subterranean cable would regard the necessary excavations.

For the easy removal of goods, the first and principal aim of the carrier is to sling them. Observers are struck by the facility and steadiness with which packages are removed from place to place by means of a crane; *e.g.*, in a ship's hold or at a railway goods-shed. I have lately introduced these hammocks to the profession for the transit of invalids by rail, because the conveyance of injured persons has been far too much left to chance, and no practical steps have been taken to remedy an inconvenience of daily occurrence. Let me direct your attention to this modified stretcher, on which is slung a small hammock. This apparatus is easily carried by two men, and is intended not only for exercising patients in the open air, but also for conveying invalids to or from a railway station, so that the patient does not quit the hammock, and consequently all changing is unnecessary. The three patients who have been conveyed from Westminster to Margate, Ringwood, and Bournemouth respectively, in hammocks by rail, have all expressed their great satisfaction; and I can personally bear witness to their testimony. With regard to the further use of hammocks, let me strongly urge that one, at any rate, should be kept at every railway station; collisions, accidents, and vivisection in various forms, are terms familiar as 'household words' to railway directors; small provision is made by them for the transit of the wounded; in any grave accident, a telegram could thus shortly produce a sufficient supply of swing-beds.

In the North of Devon I know of men with fractured legs and strangulated herniæ having been placed in a cart on a bundle of straw, and slowly conveyed sixteen miles to the Exeter Hospital. Surely this procedure is not calculated to ensure the recovery of the sufferer! Again, might not a mother's anxiety be lessened, and co-passengers' comfort be increased, by swinging babies and children during long journeys? My experience tends to prove that children very much enjoy a railway journey until fatigue sets in ; their having no suitable means of reclining transforms them into intolerable bores. Are we not more to blame than the children is not anticipating the inevitable ? In conclusion, there are some 'travelling larders and wine-cellars' who tell surgeons that a good railway jolting is to be approved of, because it shakes up their liver ; but there are also surgeons who answer that a horrid railway jolting is to be disapproved of, because it shakes up their nervous system. My sympathies are with the latter ; for there is a daily increasing class of educated men who believe that repose in travel is not so much a question of luxury as of necessity.

SUBJECT IV.

ABSTRACT OF A CLINICAL LECTURE ON THE DRESSING OF WOUNDS.

DELIVERED AT THE WESTMINSTER HOSPITAL, OCTOBER 14, 1876.

(*'British Medical Journal,' December* 30, 1876.)

GENTLEMEN,—For the last three years, as many of you know, most of the wounds received into my wards have been exposed to the air without any dressing whatever; and we may now fairly judge of the results obtained by this method.

The following is a list of the excisions and amputations performed by myself during the past two years: Excision of hip, 2; of knee, 5; of shoulder, 1; of elbow, 3; of os calcis, 1; of astragalus, 1; of cuboid bone, 4; amputation of thigh, 1; of leg, 4; Syme's, 5; Chopart's, 2; of fingers, 1; of mamma, 3—total, 33. These cases were treated by the open method, and no death resulted.

But, without asking you to pin your faith to any special style of dressing, I will briefly recall to your minds some of the ancient and modern procedures, and tell you why I prefer the method of the open treatment.

Nothing can cap the elaborate detail of ancient surgical dressings; layer after layer of compounds were heaped upon the raw surface; success was supposed to result in a direct ratio to complexity. Surgical practice ran a race with religious observances; both were highly complicated, unnecessary, and distorted the simplicity of truth.

Here is a surgical specimen.

Ambroise Paré, in 1545, writes: 'When I first came to Turin, there was there a Chirurgion farre more famous than all the rest in artificially and happily curing wounds made by gun-

D 2

shot ; wherefore 1 laboured with all diligence for two yeeres time to gain his favour and love, that so at the length, I might learne of him, what kinde of medicine that was, which he honoured with the glorious tittle of Balsame ; which was so highly esteemed by him, and so happy and successful to his patients ; yet could I not obtaine it.'

On Paré leaving Turin, he says ; ' Wherefore I went unto my Chirurgion, and told him that I could take no pleasure in living there, and that I intended forthwith to returne to Paris, and that it would neither hinder nor discredit him to teach his remedy to me, who would be so farre remote from him. When he heard this, he made no delay, but presently wished me to provide two Whelpes, 1 pound of earth-wormes, 2 pounds of oyle of Lillyes, sixe ounces of Venice turpentine, and one ounce of aqua vitæ. In my presence he boyled the Whelpes put alive into that oyle, untill the flesh came from the bones, then presently he put in the Wormes, which he had first killed in white wine, that they might so be clensed from the earthy drosse wherewith they are usually repleate, and then hee boyled them in the same oyle so long, till they became dry, and had spent all their juyce therein ; then hee strayned it through a towell without much pressing, and added turpentine to it, and lastly the aqua vitæ. Calling God to witnesse, that he had no other Balsame, wherewith to cure wounds made with gun-shot, and bring them to suppuration. Thus he sent me away as rewarded with a most pretious gift, requesting me to keepe it as a great secret, and not to reveale it to any.'

This diabolical dressing Paré recommended as a fit medicine to procure the falling away of an eschar.

Sir Astley Cooper, in his ' Lectures on Surgery,' spoke thus : It is curious to see the difference between the mode of dressing stumps now (1830) and that adopted a few years ago. The old practice was, after the adhesive plaster had been applied, to put some lint, then plaster again, after that tow, and, lastly, over the whole a cap of flannel. If a surgeon were to do this now, he would be laughed out of the operating-theatre, and very deservedly too, because he would prevent the success of the adhesive process by unduly heating the limb. All that is necessary to do is to apply three strips of adhesive plaster over the wound, and one circular piece ; if the weather be hot, to apply the spirit of wine and water lotion ; and, if it be cool, to keep the limb quiet.'

How marked has been also the change in medical practice from the old-fashioned doses of physic (containing admixtures of the most abominable compounds) to the rational treatment of disease by prevention, sanitation, and change of air and scene ! But the thought of drugs reminds me of a clinical study of much interest, viz. the dressing of wounds by internal remedies, such as the treatment of wounds of the skin by arsenic ; that of an irritable phagedænic sore by opium ; or that of a syphilitic ulcer by iodide of potassium. Although simplicity of treatment indicates your general faith, yet surgical practice should by no means degenerate into an expectant sameness.

In more recent times, the plan of irrigation, overflowing fœtid sores, gave, and continues to give, satisfaction. This overflow is practically inconvenient, because excessive ; to obviate this, Dr. Lee's steam-kettle is a good mechanism for gently applying a continuous stream of disinfecting moisture without deluge. As for poultices; I hold them in abhorrence ; they are both dirty and vulgar; warmth and moisture may be obtained by hot flannels, the steam-jet, or by warm water dressing ; but, in all cases where warm water dressings are used, the element of moisture should assuredly be constantly changed, because water is a prime factor in favouring decomposition ; the wet lint and gutta-percha sheet placed over an open sore shortly becomes impure by reason of admixture with blood and discharge, the granulating surface is swamped, and its tone enervated.

At the present time, surgical attention has been prominently focussed on the antiseptic system of dressing wounds by Professor Lister. I well know that by his method more serious operations may be performed with impunity than hitherto ; *e.g.* the free opening of joints, deep abscesses, serous cavities, &c., and that in the Royal Infirmary, Edinburgh, admirable have been the results in his own hands ; but, with all due deference to so high an authority, I cannot but think that his hobby has been ridden hard, and that a tedious expenditure of energy occurs in the Edinburgh school. Will the majority of English surgeons admit that the antiseptic system is necessary for tenotomy ? or for the operations of minor surgery ?

By so pointedly directing attention to the dressing of

wounds, the minds of surgeons are in danger of becoming narrow; for the wound itself is but a single element in a surgical case, and in many instances the amount of suppuration is of no vital import; the bugbear pyæmia is to be combated by sanitation, only one condition of which is supplied by the dressings.

If I may liken a surgical case to a domestic establishment, Lister's treatment of wounds is represented by a householder who disinfects his drain; a most wholesome practice, but not so absolutely essential as to justify the stigma that an ordinary well-trapped sewer is impure and untrustworthy. In the same way as many establishments are well regulated and successfully conducted without such artificial remedies, so also many cases in surgery are to be treated satisfactorily without antiseptics. My own experience at this hospital supports this view.

Professor Lister has by his talent, given surgeons a grand ally in performing operations where death is to be apprehended from exhaustive discharge, or from the results of interference with synovial or serous cavities. In ordinary surgical work, I question the necessity for the antiseptic system; 'the game is really not worth the candle;' and, until I learn that the open treatment of wounds is unsuccessful, I shall continue it, for the following reasons :—

1. Our results are equally as good as by the antiseptic system; no death having occurred from pyæmia or exhaustive discharge.

2. Trouble and expense are reduced to a minimum.

3. The fullest opportunity is granted to students for clinical observation; on the antiseptic system, the wound is but seldom and briefly exposed.

4. All nervous apprehension from the indiscreet removal of, and the painful repetition of, dressings is done away with.

5. The process of healing by scabbing is solicited.

6. Nature is duly accredited with her share in the performance; and a host of lotions and ointments are dismissed as plagiarists.

I have been in the habit of comparing surgical dressing to the dress of individuals; the simplest suits best; the elaborate appears most ridiculous; and I strongly suspect that as much of the drapery displayed now-a-days is unnecessary and expensive, so also are many of the contents of a dresser's panier. By divesting ourselves and our wounds of excessive dress and dressings, we prepare both for repose and the reparative process.

I have said this much from the firm conviction that in surgical dressings much that is done had better be left undone; dressings are not to eclipse the beneficent repair of Nature, but need only (if used at all) be neat, pure, and simple to be effective.

SUBJECT V.

CLINICAL LECTURE ON RECTAL EXAMINATIONS.

DELIVERED MARCH 31, 1879.

('*British Medical Journal*,' *July* 21, 1877.)

GENTLEMEN,—Within the past month we have had a series of cases illustrating features of surgical interest in connection with the rectum. I shall first direct your attention to some anatomical facts, and then offer clinical remarks on the treatment of fistula *in recto*, fissured anus, pendulous skin-tags around the anus, piles, and lastly, on the introduction of the surgeon's hand and arm *per rectum* for left renal examination. Hilton, in his classic work, 'On Rest and Pain,' draws attention to a white line of demarcation between the external and internal sphincters ; the former I have been in the habit of describing as the sphincter ani, the latter as the sphincter recti. They both act as janitors ; but the anal sphincter specially guards against intrusion from without, whilst the rectal sphincter, on receiving the order from his superior officer, carries out the final expulsion of fæces. I apprehend that nature has erected Houston's folds on very much the same principle that barriers are erected in public places, viz. to prevent over-crowding, and to give supportive protection to and from ' the residual element.' The common iliac arteries may be felt pulsating right and left of the rectal column ; I have lately utilised this anatomical fact for the restraint of hæmorrhage during amputation at the hip joint; the position of the internal iliac arteries also with regard to the pelvic wall suggests that the mechanical pressure of the fœtal head and body during parturition may assist in checking flooding in cases of placenta prævia. The healthy anus is an even, tight, depressed ring of tissue, dilating easily to a diameter of one inch and a quarter, sensitively resenting intrusion from without both of solids and liquid, whereas the rectum may be distended to a diameter of four inches, and is tolerant of manipulation. Here you may examine the simple yet very effective speculum for the rectum that I am in the habit of using; it is preferable to Hilton's window speculum,

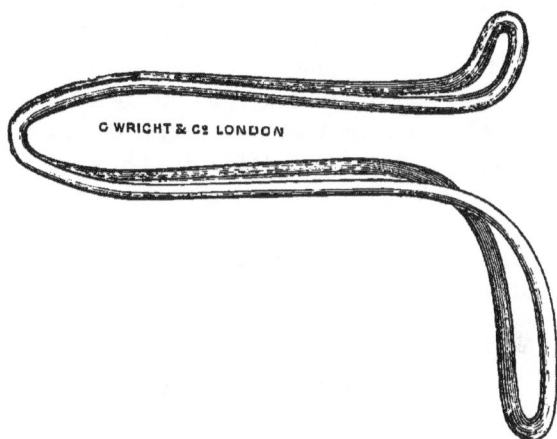

FIG. 14.

because it is larger, cheaper, and exhibits the whole of the rectal walls at a glance; the instrument itself occupying the minimum of space.

Before speaking on the treatment of some diseases of the rectum and anus, let me remind you that the public recognises specialities in our profession; your intelligent Britisher swallows advertised sherry, homœopathic surgeons, throat, skin, and anus-doctors; the last bracket is infelicitously equivocal, for one meaning of 'anus' is *an old woman*. General hospital surgeons do not receive the same amount of patronage that specialists do; patients will ask highly qualified men after specialists. Sir Thomas Watson has been asked to recommend a good general physician. A stockbroker lately asked me, ' Who is a good man for the gravel ?' I replied, Scotchmanlike, ' Who is a good broker for an investment in Russian bonds ? ' I trust that the time is not far distant when the Local Government Board shall recognise substantially the grand service done by general hospital staffs to the community, instead of perpetuating the present competition between the great unpaid general hospital men and the overpaid specialists. I have so far digressed to direct your ambition towards becoming 'good all-round surgeons,' and in order that you may study and practise surgery as an integral and not a fractional quantity.

Let me now draw attention to the common complaint called fistula *in recto*. In your text-books, and even in Bryant's 'Surgery,' this complaint is erroneously called fistula *in ano* ; the anus is but a linear entity, length without breadth, the mouth of the sewer. If two apertures exist, one is in the skin around the anus, the other in the lower part of the rectum. In all operations on the rectum, attend to the following hints : 1. That the bladder be empty; 2. That the banks of the sewage-canal be well washed previously with tepid soap and water; 3. That your own hand be oiled, the nails cut short, and the semilunar folds around the nails filled with soapy smears; 4. That suitable lavatory accommodation (with disinfectants) be ready for your own ablution. Divide the fistulous track accurately; retraction of the sphincters results; the floor of the sinus is free from friction, and granulations adhere from the floor towards the circumference of the anus. Administer opium *per rectum* or by the mouth as occasion seems fit, so as to quiet the intestinal tract. Nitrous oxide gas may well be used for these short operations, because sickness is a very aggravating occurrence as a sequence. In fissured anus, where exquisite sensibility occurs, and the patient carries the aspect that Professor Miller indicated as *mens conscia recti*, divide the nervous filaments in the sulcus, as explained by Hilton.

In considering the treatment of hæmorrhoids, bear in mind this standard rule, that whatever you remove as *bonâ fide* integument may safely be done by a curved pair of scissors; whatever is removed as *bonâ fide* mucous membrane must be done by clamp and cautery, or by ligature. In the out-patient department, I constantly cut off cutaneous tags around the anus with scissors, and treated the owners of them as out-patients; but never did once venture so to handle hæmorrhoids. Patients with piles require careful rest and attention prior to operating. Mr. Henry Smith has introduced a pair of forceps for holding these slippery customers, and then burns them off; my own experience inclines me to hold that the whole of the curative credit is due to the cautery. At this hospital we use a much stronger clamp than Smith's, and more after the St. George's

Hospital pattern, because the grip is much firmer and cannot slip. (The instrument is on the operating-table.) The cautery I now use is this very ingenious and practical platinum cautery invented by Dr. Paquelin; after heating the platinum-tube over a spirit-lamp, the vapour of benzoline driven over this heated surface maintains its temperature as a cauterising agent. The late Mr. Bruce devised an arrangement for utilising common gas for cautery purposes, but benzoline is more portable than gas. No modern invention has succeeded so well in neutralising the apparent brutality of the hot iron as this of M. le Docteur Paquelin. These rectal wounds must all be treated by consummate cleanliness, and require no other dressing. You will find the steam-spray very grateful, and a soft hog's bristle brush a useful adjunct.

I have yet time to bring before your notice a case of renal phthisis, in which I introduced my right hand and forearm *per rectum*, so as to examine the left kidney for suspected calculus. T. B., aged 35, a painter, was admitted into Mark ward on January 17, 1877. Mr. Peacock, of Lincoln (under whose care T. B. was originally), states that the man had been treated for lumbago, and subsequently required baths at frequent intervals. His wife, to whom he had always been reticent, determined to know why he required baths (apparently an unusual demand on the part of the British workman), and she discovered that his 'purse' and left testicle were swollen. In 1876, hæmaturia, especially at night, occurred off and on with symptoms of cystitis. Mr. Peacock treated him with the bicarbonate of potash and henbane, perchloride of iron, and triticum repens. On admission, the patient stated that he was supposed to be suffering from stone, either in the kidney or bladder. In 1867, he fell eighty feet, while painting the roof of St. Andrew's Church, Derby. In October 1874, after exposure to cold, he complained of pain in his left loin; he had a fall in his shop at this time, and so frequent were these falls in his career that I nicknamed him 'Tumbling Tom.' In 1876, he suffered from hypogastric pain, especially on passing urine, with an excruciatingly crushing pain in his left testicle. The catheter was passed for him now on fourteen different occasions; blood in his urine appeared, but gave him relief. For the last four months he had been in the Lincoln Hospital, and the treatment pursued had been directed to his genito-urinary tract. His father died, aged 64, of gravel; and the patient had been married for fourteen years without any family. His nervous, respiratory, circulatory, integumentary, and alimentary systems were apparently normal.

On admission, he complained of the gravest torment in his loins, especially on his left side; this agonising pain was persistent, and passed forward to the hypogastrium. His left circumflex ilii vein was dilated, and so were the venous radicles around the upper and outer side of the left thigh. The pain was referred down the branches of the left external cutaneous, and both genito-crural nerves. His left testicle was supremely irritable and sensitive; and a distinct induration was to be felt at each globus minor of the epididymis. Neither testicle was retracted. He wet his bed every night, and had a frequent desire to micturate by day; the urine was very faintly acid or quite neutral, rapidly becoming alkaline, of specific gravity 1030; it contained blood, pus, and mucus, and stank (as he said) like the smell of a dead body. No fragments of stone had been seen. Albumen was inconstant. He had œdema of the ankles after standing. He was ordered a grain of opium twice a day, daily four ounces of port, four pints of milk, two pints of beef-tea (double strength), two eggs and toast and water. No urethral stricture nor calculus in the bladder was found; and feeling that there was considerable diagnostic doubt, after a consultation with my colleague Dr. Basham, I proposed to introduce my hand *per rectum*, and ascertain whether or not a renal calculus could be felt by this manipulation.

January 30, 1877.—He first had a free enema of warm soap and water. Under the influence of chloroform, I injected about half a pint of sweet oil into the rectum; cut my nails, soaped my right hand, and by a screw-like movement inserted my hand as a cone into the rectum; the sphincter ani was felt to split, and slight oozing of blood followed. The promontory of the sacrum was felt, the common iliacs, and the aorta as a pulsating column. The sigmoid flexure was then carefully traversed; the transverse processes of the lower lumbar vertebræ were felt; and, lastly, the lower fourth of the kidney as a rounded tumour. Beyond this my hand would not pass, but impinged upon the velvety wall of the colon. The capacity for using the hand as an independent tactile agent was not great, yet as a counter-resistant with external manipulation it was capable of receiving tactile impressions. On the inferior surface of the kidney were two elongated nodules, which did not pulsate. The post mortem examination proved these to be a knuckle of contracted large intestine (descending colon). The hand was gently and gradually allowed to be forced down by the colon and rectum ; it was slightly smeared with blood, and somewhat cramped, reminding one of having performed version in a parturient woman. The anus collapsed to about the diameter of a crown-piece ; and the whole procedure occupied ten minutes. Before removing my hand, the posterior wall of the bladder was supported by the hand : the sound was passed *per urethram*, but no stone was detected. The man was put into a warm bed, and a grain of opium given twice a day.

January 31.—Two grains of opium were given extra, because the man had such grievous pain in his back and left loin. He said that his back must break. The urine was slightly tinged with blood. No fæces passed.—4 P.M. Sickness (of food) came on, with profuse perspiration. The two grains of opium gave the man relief by making him drowsy. He had ice to suck.

February 1.—Since 9 A.M. the sickness had ceased, but nausea continued : the pain was not so severe ; he took his fluid nourishment eagerly. At 4 P.M. sickness recommenced, and ceased at 8 P.M.

February 2.—He slept a little : the pain was not so severe : he could now lie on his back, whereas previously he was compelled to lie on his right side.

February 3.—Motion was passed involuntarily.

February 5.—The anus was of the size of a half-crown piece ; fæces were passed involuntarily ; the pain was lessening : the urine was clear, slightly alkaline, and of a faint sickening smell. Dr. Basham saw the man ; he considered the case as one of pyelitis, and advised pareira brava.

February 6.—Great pain was complained of over the left crest of the ilium : the man's tongue was clean, and he passed a good night : the testis was retracted. He was ordered to have fish and eight eggs daily.

February 11.—The symptoms continued the same ; some

days the urine was clear and healthy, at others turbulent and bloody (neutral to test-paper). I thought it right to send for his wife to-day, because he was not improving, and also to cut off his opium, as the drug might mask his progress. His right testis was especially sore to-day; neither was retracted.

February 17.—The symptoms continued the same, but on the whole the man was weaker, and his face was pinched.

February 24.—Continence of fæces was regained for the first time since January 30.

February 27, 2.30 P.M.—He complained of very great pain over the lower part of his belly, corresponding to the hypogastric region; sloughy shreds passed: and he fancied he passed urine *per anum.* At 3 P.M. he was partially dressed, and sat up for twenty minutes; then he had to go to bed again, because he felt tired: the pain in the back and belly gradually increased, and he had presentiment of death at 5 P.M.

February 28.—He had excruciating pain in his back: he begged to be placed on his face, which was done, and he died in this posture at 3.20 A.M.

'Post Mortem Examination' on February 28, at 3.30 P.M. The body was well nourished. Post mortem rigidity was present. The head was not examined. The thoracic viscera were healthy, excepting emphysema at the edges of both lungs. On opening his abdomen, there was no evidence of any peritonitis; the small intestine was irregularly and abruptly contracted for about two and a half inches at the commencement of the ileum, and the large intestine was also contracted to a minimum at the junction of the ascending with the transverse colon, and also from the descending colon to the rectum; the bowel was not adherent to either kidney.—*Kidneys :* The right was hypertrophied, and had two small serous cysts on its periphery; its envelope stripped off easily, and on section healthy structure was exposed : its weight was six ounces.—*Left Kidney :* Only a fibrous capsule remained, and this was nodulated as if twenty small white marbles had been squeezed when in a soft state into the fibrous bag : each tubercular deposit was pultaceous, and like white soft mortar : each marble had its fibrous investment, which led towards the pelvis of the kidney. At the lower part of the organ a collection of pus (four ounces) was found, and after the matter had been washed out this remnant of a kidney weighed four ounces: it was contracted to half the normal size: the pus was faintly fetid.—*Ureters:* The right ureter was a double one, leading up to two pelves: these two ureters joined after leaving the two pelves at a distance of three inches from the kidney: their structure was healthy. The left ureter was shortened, stiffened, and of variable size from the more or less tuberculous deposit in and around its walls: at the point where it crossed the left common iliac artery its coat had given way, pus and tuberculous *débris* and blood had gravitated downwards behind the rectal column, and had reopened into the rectum by a small ragged opening on its posterior wall. The apertures of the ureters in the bladder were dilated.—*Suprarenal capsules:* The right was healthy, the left much atrophied.—*Bladder:* The coats were thin, but otherwise normal: no deposit was noticed.—*Testes:* Both testes were normal in size, and glandular structure unaltered to the naked eye: no spermatozoa were found; the globus minor of each epididymis held a yellow tuberculous patch of the shape of a pea; and both vasa deferentia and vesiculæ seminales were choked with yellowish-green tuberculous paint: this stuff could be squeezed along the vasa into the urethra. The urethra was normal: there was no stricture. The upper three-fourths of the rectum, the sigmoid flexure, and descending colon, were healthy. The anus was patent to the size of a shilling, and the sphincteric structures were thickened, apparently from organised lymph; some fæcal matters were found in the rectum.

Mr. Thomas Smith has carefully recorded six cases of this kind in vol. viii. page 95, of the 'St. Bartholomew's Hospital Reports,' and his simile is apt in placing the disease on a parallel with phthisis pulmonalis. Frequent micturition, hæmaturia, by night (*i.e.* after the day's fatigue), intermittency of subjective symptoms, pus, and mucus in the urine closely pictured the features of renal or vesical calculus; the rectal aperture explained the man's sensations on the day before his death; while the starting points of this train of pathological results may be assumed to have been three in number, viz. from the left kidney and the two epididymides. The man's sterility is obvious. The engorgement of the left circumflex ilii vein is accounted for by the direct pressure on the left ascending lumbar veins. The relief gained by passing the catheter was probably mental, or divertent; but the surgeon's power for good in phthisis renalis (so far as drugs are concerned) seems to be of a most limited nature.

The introduction of the whole hand into the rectum was practised by Mr. Maunder in the year 1866, for pelvic diagnosis and for strictured large bowel: in Holden's ' Landmarks' (1876), page 70, there are some remarks by Mr. Walsham, demonstrator of anatomy at St. Bartholomew's Hospital, on rectal palpation. He has introduced his hand into the rectum for the diagnosis of strictured sigmoid flexure with doubtful result. Mr. Walsham's hand measures in circumference about seven and a quarter inches; my own measures eight inches. I consider that the introduction of four fingers into the rectum is a valuable method of diagnosis, say, for stone, or uterine ailments, or for strictured intestine; that the insertion of the whole hand and forearm is a possible but severe procedure; that occasionally it may aid diagnosis (in my own case it enabled me to exclude stone in the bladder or kidney or ureter from consideration). This mode of examination can never be available for general practice, because the information thereby gained is not equivalent to the incurred distress; and let me add, that once on the post mortem table my hand ruptured the superior portion of the rectum. The pathologist's hand through the rectal tube has been utilised in post mortem examinations for the abstraction of the kidneys and other small viscera: I mention this only to express my regret that any such underhand examination has been necessitated, by reason of friends negativing an essential but disagreeable duty of our profession in the pursuit of truth.

SUBJECT VI.

ABSTRACT OF A CLINICAL LECTURE ON RESECTION OF THE KNEE-JOINT IN YOUNG SUBJECTS.

DELIVERED AT THE WESTMINSTER HOSPITAL, JUNE 12, 1876.

(' *Medical Examiner*,' July 27, 1876.)

GENTLEMEN,—There is now in Mark ward a boy, æt. 14, on whom the operation of removing the whole of the left knee-joint was performed May 9, 1876 ; he left his bed on June 12, 1876 (thirty-four days subsequent to the operation), with a stiff ankylosed limb. This case illustrates a chain of events which successively rivet our attention—

1st. On the reasons why the operation was undertaken.

2nd. On the method according to which it was performed.

3rd. On the after-treatment.

4th. On the future condition of the stiff leg.

1st. This happy-dispositioned boy acted as my guide in the spring-time, hobbling about Westminster on a crutch, with his left leg swinging in mid-air. He told me how his bad leg inconvenienced him, preventing him from gaining a livelihood, because employers objected to the look of the crutch, and mistrusted his capabilities. The joint was evidently becoming firmly ankylosed at an angle of 70° ; a sinus existed on the inner side of the knee, through which exuded a watery pus. His disease was originally lit up by a fall on the gravel, and his deformity had been gradually increasing for eleven years. The boy, with proper pluck, requested to be operated upon ; and he was admitted on May 6, 1876.

2nd. Esmarch's bandage for restraining hæmorrhage is much to be commended for this operation. Make a simple transverse cut across the line of articulation, and do not remove any skin. Saw carefully through seven-eighths of the lower end

E

of femur, and fracture the other one-eighth, so as to avoid any injury to the posterior ligament of the knee-joint or the popliteal artery. Saw through the upper part of tibia, and accurately adjust these bones until they co-adapt. I have latterly pruned the sawn bones with a strong knife, and for the following reasons. In the vegetable world a sawn surface is apt to canker, and the expert gardener freshens his saw-cut by the subsequent application of the chisel or knife. I have thought also that the surgical law of union in respect to incised and lacerated wounds of the soft parts may with equal justice apply to bone: that is, as an incised wound heals more quickly than a lacerated one, so cut surfaces of bone may more readily unite than sawn ends; therefore, first wash your saw-cuts with a stream of carbolised water, to get rid of *débris*, and then make fresh the femur and tibia by a keen instrument, ere resting them finally together for con-solidation. Bring the wound together by bead sutures, using the double wire needle, and lastly apply the splint here shown, defended by india-rubber padding, on the anterior aspect of the thigh and leg.

FIG. 15.

I have completely discarded the pos-terior splint, because it is not so cleanly for discharges, and because it so frequently produces a sore heel. Let the limb rest on a plane surface, and anticipate start-ings by a moderately heavy sandbag.

3rd. The after-treatment is the same as for any compound fracture ; keep the parts at rest, dry and clean beneath a cradle ; use no dressings; and swing the limb as soon as convenient. Above all, let the patient have a generous meat diet.

4th. After ankylosis of an osseous kind has resulted, the patient must wear a high boot, and so arrange the artificial sole that the leg can readily clear itself off the ground in progression. Sinuses not infrequently endure for some months after union has taken place ; they are best left alone. The legs do not grow symmetrically, because the epiphyses of the condyloid end of the femur and head of the tibia have been wholly or partially removed. Nature's architectural plan has been interfered with, and the result is a real and apparent increase of the deformity. This experience does not apply to the fully grown, and need not deter any surgeon from resecting a knee-joint in a young person.

At the Royal Infirmary, Edinburgh, my late teacher, Mr. Syme, used to state that excision of the knee was an unjustifiable operation; but extended observations cannot fail to persuade surgeons that this operation is not only justifiable, but ranks in the van of the beneficent deeds of our art.

It will be seen from this table that I have had the good luck of treating twenty excision cases of the knee-joint in succession without any fatal result. I cannot reasonably expect such good fortune to continue. The open treatment of the wound has been strictly adhered to, without any special antiseptic dressing; some credit may justly be claimed by the surgeon for his watchful care, but much also must be given to the skill and obedience shown in these severe operations by nurses, trained under my own regulations; and to these faithful allies at Westminster I offer my heartfelt thanks.

The accompanying table records tersely my own experience.

Table of Resections of the Knee-joint.

Mr. RICHARD DAVY.

No.	Sex	Age	Admission	Year	Day of Operation	Discharged	Complaint	Nature of Operation	Days in Hospital	Days after Operation	Result
1	Boy	13	May 23	1871	May 30	Aug. 15	Disease of right knee. Genu valgum	Excision of internal condyle of femur	84	77	Ankylosis, Fibrous.
2	Girl	2	April 15	1872	May 14	July 2	Ankylosis. Left. Deformity	Excision of bony wedge	78	48	Osseous,
3	Girl	1	Feb. 24	1873	March 4	July 23	Ankylosis. Right. Deformity	Ditto	149	110	Ditto
4	Boy	8	July 21	1873	July 28	Oct. 18	Suppuration of right knee-joint	Excision of right knee-joint	89	82	Ditto
5	Boy	4	Jan. 1	1874	Jan. 13	April 29	Suppuration of right knee-joint	Ditto	119	115	Ditto
6	Boy	3	March 2	1874	March 24	May 27	R. Fibrous ankylosis.	Ditto	86	64	Ditto
7	Boy	8	May 16	1874	May 19	Aug. 5	Ankylosis. Deformity after resection. L.	Excision of previously excised knee-joint	81	78	Ditto
8	Girl	11	April 14	1875	May 18	Aug. 2	Chronic synovitis, with deformity. R.	Excision	110	76	Ditto
9	Boy	14	May 25	1875	June 1	Oct. 28	Genu valgum. R.	Ditto	162	156	Ditto
10	Girl	4	Nov. 2	1875	Nov. 9	Feb. 20	Strumous degeneration and dislocation. R.	Ditto	110	103	Ditto

No.	Sex	Age	Admission	Year	Day of Operation	Discharged	Complaint	Nature of Operation	Days in Hospital	Days after Operation	Result
11	Boy	14	May 6	1876	May 9	July 17	Deformity and ankylosis of left knee-joint	Excision of the whole joint	72	69	Osseous Ankylosis
12	Girl	11	March 6	1878	March 13	May 28	Synovitis. Right	Excision .	76	69	Ditto
13	Girl	9	Feb. 15	1878	Feb. 26	May 16	Acute synovitis. R.	Ditto	92	81	Fibrous Ankylosis
14	Man	24	March 16	1878	March 19	May 9	Rectangular deformity. Left	Ditto	54	51	Amputation. March 21
15	Girl	15	June 20	1878	July 2	Oct. 2	Rectangular deformity. R.	Wedge freely cut out	104	92	Osseous Ankylosis
16	Boy	5	Feb. 7	1879	Feb. 11	May 20	Dislocation. Old R. Synovitis. Necrosis.	Wedge cut out	102	98	Ditto
17	Boy	6	July 28	1879	July 29	Nov. 15	Synovitis. Right knee	Excision	109	108	Ditto
18	Woman	30	May 16	1879	May 27	July 29	Synovitis. L.	Ditto	63	52	Ditto
19	Boy	6	Jan. 6	1880	Jan. 20	June 7	Ankylosis. L. knee	Ditto	152	138	Ditto. (Admitted twice)
20	Girl	11	Jan. 14	1880	Jan. 27	March 23	Chronic synovitis. R.	Ditto	68	55	Fibrous Ankylosis

SUBJECT VII.

ABSTRACT OF A CLINICAL LECTURE ON AMPUTATION THROUGH THE KNEE-JOINT.

DELIVERED AT WESTMINSTER HOSPITAL, NOVEMBER 6, 1877.

(*'Medical Examiner,' December* 13, 1877.)

GENTLEMEN,—I have the opportunity to-day of directing your attention to four cases of amputation through the knee-joint, which have been treated in our wards.

Professor Lister in his essay (Holmes' ' System of Surgery ') devotes twelve lines to the consideration of this amputation, stating that ' amputation at the knee-joint is an operation which has not met with much favour. ' Professor Syme, in his ' System of Surgery,' does not mention it ; Professor Erichsen, in his work on surgery, notices it favourably. I have no hesitation in strongly upholding this operation, and will briefly state *why* and *how* it is to be performed, *in what cases*, and *with what results :*—

1. *Reasons for selection.*—In amputating through a joint there is no division whatever of bone ; Nature has decided that question, and has focussed important structures, such as artery, vein, nerve at the knee, so as to centralise interference.

The minimum of tissue is injured, viz., ligaments, skin, fascia, and tendons.

The naturally smooth and obtuse end of the femur not only presents an admirable base for the stump, but also receives the superimposed weight in the horizontal plane.. The patellar skin is tolerant of pressure.

The leverage of the triple group of femoral muscles is maintained.

2. *Method of operating.*—In young and healthy subjects I prefer the long square anterior flap and transverse posterior ;

Table of Amputations through the Knee-joint.

N°.	Name	Age	Sex	Occupation	Admission	Disease	History	Date of Operation	Discharge	Days in Hospital after the operation	Result
1	M. M.	60	M.	Drayman	July 17, 1876	Fr. tibia and fibula, and patella. Ulcer on leg for 6 years	From a fall of 14 feet	July 18, 1876	Oct. 15, 1876	89	Most useful stump. Artificial leg
2	A. T.	14	F.	Schoolgirl	Nov. 7, 1876	Flail leg. Atrophy. Talipes equino-varus	Fit when 2 years old	Nov. 14, 1876	Mar. 1, 1877	106	Most useful stump. Artificial leg
3	P. L.	56	M.	Labourer	Dec. 26, 1877	Potts' fracture. Gangrene	Fall when tipsy	Jan. 7, 1877	Jan. 27, 1877	20	Death
4	G. S.	4	F.	Schoolgirl	Oct. 30, 1877	Talipes equino-varus. Infantile paralysis of leg	Since commencing to walk	Nov. 6, 1877	Jan. 5, 1878	60	Excellent recovery
5	H. G.	19	F.	Dressmaker	Jan. 1, 1880	Useless right leg. Sores	For 16 years	Jan. 27, 1880	May 16, 1880	110	Most useful stump. Artificial leg

in old persons a long posterior flap, because the posterior seg-
ment carries the greater supply of blood.   Your object is to
render the face of the stump scarless, and your ingenuity will
suggest individual application.   Let me state this surgical
axiom, that in planning flaps, wherever you have reason
for apprehending defective nutrition, base your procedure on
surgical anatomy, conserving vascularity.   Long flaps are prone
to slough, however closely you may hug the bones, and you
must, in fair decision, yield the natural drainage of a long anterior flap to the artificial drainage tube sometimes necessary
with a long posterior flap. Open the joint anteriorly by dividing
the ligamentum patellæ and capsule ; sever the sartorius, gracilis,
and lateral ligaments ; sedulously avoid injuring the condyloid
cartilage when cutting the crucial bands ; and, lastly, divide the
hamstrings, artery, vein, and nerve.   I insist strongly on your
not meddling with the femoral articular cartilage by scraping,
sawing, or cutting.   It is wholly unnecessary, and converts
amputation through the knee-joint into that of the thigh.

Twice in my lifetime have I seen two different surgeons compelled to saw off the condyles of the femur, by reason of the great
retraction that occurs in the flaps, after having removed the leg
and semi-lunar cartilages.   Now, why does this great retraction
occur ?   Skin is proverbially elastic, especially where flexion
and extension occurs ; but I apprehend that the chief factor
consists in the attachment of the fascia lata to the prominent
lines around the knee-joint, tuberosities of the tibia, and head
of the fibula.   Unrestrained muscular contraction follows as a
result of the dissection as surely as tight trowsers slip up with
broken understraps.   Make, then, due allowance for this fact.
Of the two err on the side of redundancy ; for while no surgeon
can justify too short flaps, there are many arguments in favour
of too long ones, as, for example, the lessening of tension, the
insurance against subsequent loss of tissue, and the natural
disposition of cicatricial tissue and skin by contraction to adjust
themselves to altered relationships.

Should the patella be left or not ?   Most decidedly leave it
alone.   You gain nothing by removing it in a healthy joint, at
any rate, and you add to the severity of the operation.   In the
case of M. M., who was the subject of an antecedent fractured
patella, with an intervening link of two inches of fibrous tissue,

I removed it, as a pathological specimen, at the man's request. You will only take it away should disease or a smash compel you so to do.

The surgical instruments that you need are Esmarch's bandage, a good knife, artery and dissecting forceps, operating needle, and ligatures.

All my cases have been subsequently treated without any superficial dressing whatever, the wound being allowed to drain itself into oakum and heal.

3. In what cases should this operation be employed? In severe injuries at or near the knee-joint, we may draw the surgical line at this articulation; in wounds involving the popliteal artery and vein; in cases of gangrene from obstructed popliteal, or anterior or posterior tibial arteries; in malignant tumours of the tibia or fibula; in cases of flail-like limbs, due to infantile paralysis, where other treatment has failed; in short, amputate at this joint whenever the structures distal to the joint are useless or offending.

4. With what results is this operation attended? Next to Syme's stump this is the best: smooth, firm, tolerant of pressure, scarless, and capable of receiving the socket of an artificial limb admirably. The operation, when skilfully completed, is not attended with much shock or danger, and the single death that resulted in our hospital, where amputation through the knee-joint was performed, was due more to constitutional and accidental than operative causes.

SUBJECT VIII.

ABSTRACT OF CLINICAL LECTURE ON PUNCTURE OF THE BLADDER PER RECTUM.

DELIVERED OCTOBER 15, 1874.

(*'British Medical Journal,' December* 4, 1875.)

GENTLEMEN,—This operation is a most valuable one for cases of retention of urine, where any physical obstruction prevents the introduction of a catheter into the bladder, after fair trial on the part of a competent surgeon.   It was originally introduced into practice by Mr. Cock, consulting surgeon to Guy's Hospital;

FIG. 16.

and his wide experience coincides with my own comparatively limited observation, that there is scarcely any surgical operation so devoid of danger as puncture *per rectum.*   The necessary detail of the old operation is now demonstrated to you on this dead body; and the instruments are the same as generally used, viz., a solid trocar and cylindrical cannula, the latter being retained in the bladder by tapes.

In 1870, being impressed with the disadvantages of retaining for any length of time an unyielding tube *per rectum et vesicam,* I introduced the instrument (fig. 16), in which the steel slotted cylinder is outside, and acts as a perforator, carrying the elastic catheter as an inside passenger. Fig. 17 shows the catheter withdrawn, also the reduplication of the tip of the catheter, by means of which it is retained in the bladder. The patient is placed after the operation on a mattress having a circular hole corresponding to the buttocks; the end of the catheter falls through this hole, and conducts the urine into a receiving vessel on the floor below. Fæces also are thus passed without moving the patient.

I have yet further applied this principle of giving the urethra rest to those cases of obstinate perineal fistulæ that are *not dependent* upon an obstructed urethra for their persistence. I am of opinion that a simple operation like puncture *per rectum* is more effectual and easy than periodic introductions of a catheter; the object being in both cases to avert the irritating urine from the damaged urethra and fistulæ, especially so

FIG. 17.

because in some cases men have not the power of retaining their urine; and it is impossible by the most skilful catheterisation to avoid some disturbance to the urethra, and some extravasation of urine over the exposed surface of it and the fistulæ.

I have been in the habit of comparing the *rationale* of puncture *per rectum* in fistulous cases to the method pursued by workmen when repairing the damaged banks of a running stream. Their first object is to completely divert the stream from the seat of their repairs, so that one day's wash shall not outweigh one day's work; and this argument applies with double pathological force to the irritating urine as compared with the pure water. But a moment's reflection makes a surgeon see how inapplicable Cock's method would be for opening an empty bladder; and, at the risk of the charge of superfluity, I venture to make you acquainted with a new instrument, and a method of operating on this much-operated-on human perinæum. The instrument consists of a staff, on which slides a silver cylinder, and a removable handle. (Fig. 18.) The silver tube runs from the bulge at the right end of the staff to the commencement of the curve on the left. The thin stem underneath the handle is for the attachment of a retentive catheter.

FIG. 18.

New method of operating.—First introduce the staff into the bladder; turn the tip of the staff towards the base of the bladder and rectum; feel for the point with the finger in the rectum, and carefully cut the recto-vesical tissues until you can pass the staff out at the anus. Next unscrew the handle and affix the stem of one of Napier's retentive catheters to the thin end of the staff; soap well the india-rubber; then, holding the silver tube firmly in one hand, draw the catheter (excepting its bell-shaped end) completely into the silver tube by withdrawing the staff at the anus. Lastly, draw back the silver tube and catheter through the urethra into the bladder; push in the catheter, and free the silver tube by also withdrawing it through the rectum out at the anus. The campanulate end of the catheter unfolds itself in the bladder, and its stem loosely hangs at the anus.

I will now briefly narrate two cases of perineal fistulæ that have been operated on by this method in the Westminster Hospital :—

CASE I.—W. H., 35, slater, in 1859 fell on his perinæum across a door. Extravasation of urine resulted, and he was treated in the Yarmouth Hospital by free incisions. On leaving the urine passed through a fistulous opening in perinæo.

In 1862, a surgeon, while attempting to pass an instrument, pushed the catheter into his rectum; after that the patient thought operative procedure had better be discontinued.

In 1869, on rearing a heavy ladder, extravasation of urine recurred, and Mr. Teevan admitted him into the West London Hospital. Mr. Teevan on three occasions divided his stricture without any guide, and lastly, on account of the cartilaginous state of the parts, opened the urethra from the stricture to the bladder; No. 12 was tied in, and in five weeks he left.

In 1871 Mr. Barnard Holt split him, because two catheters had been broken and many bent in his urethra by surgeons.

On March 18, 1873, he was admitted into Luke Ward. The perinæum was a mass of cicatrices and puckerings. There were five fistulæ, one cartilaginous, at the central point of the perinæum. He had complete incontinence of urine. At the point where the penis joins the scrotum was an urethral dilatation holding three and a half drachms of urine, not albuminous. No. 10 sound readily entered the bladder. The man's general condition was good and his spirits plucky.

On March 25, 1873, I performed the operation of opening his bladder by the new method. A retentive catheter, as shown in fig. 17, was inserted, and kept in for six weeks. He most faithfully lay on his back on a circular air-cushion, urine and fæces passing through the circular hole in the cushion. After six weeks of this endurance four of the five fistulæ were firmly healed; the grand central perineal fistula was not. A urinal was easily adapted to this single outlet, and he left the hospital on June 6, 1873, well satisfied with the result. The recto-vesical opening was completely closed.

CASE II.—E. P., 24, porter, was admitted into Luke Ward on July 24, 1874. Five months previously he had gonorrhœa; a month subsequently a lump formed in the perinæum, and a urinary fistula resulted. The man was of strumous aspect. The urine was not albuminous. A No. 10 sound could be passed. July 26.—I injected the fistulous course with compound tincture of iodine, but with no result. August 21, 1874.—The same operation was performed, and the same treatment pursued, as in case I., until September 15, 1874. The house surgeon injected the fistulous track with tincture of iodine, which set up urinary infiltration, and the retentive catheter was withdrawn.

My colleague Mr. James Keene (who had charge of my beds during the autumn vacation) repunctured his bladder per rectum, and watched the case with much assiduity; but the extravasation continued. On my return, October 5, the patient's condition was critical, the inflammatory process extending up the abdominal wall to the umbilicus. As the urine came freely through a sloughy hole in his perinæum I at once withdrew the catheter from his bladder. He was well supported with nutritious food, stimulants, and opium. On October 14, his perinæum being very sloughy, I laid the parts freely open; he lost much blood during the operation, and died collapsed the same evening.

Post mortem examination.—The friends gave permission for the bladder only to be examined. The perinæal structures were a confused mass of sloughy shreds; there was recent lymph thrown out over the vesical peritoneum. The interior of the bladder was apparently healthy; the first puncture had completely healed; the second admitted but a small pin; the sloughy mass extended along the scrotum upwards over the whole of the abdominal wall below the umbilicus.

The conclusions arrived at from these two cases are the following:—Case I.—Puncture *per rectum* is insufficient *per se* to cause the obliteration of an intractable cartilaginous fistula, but affords the means of occluding lesser and more recent fistulæ in perinæo. A retentive soft catheter may be retained in the bladder for six weeks, and the wound made in the rectovesical tissues will subsequently heal. Case II.—The injection of iodine through a fistulous track in the perinæum of a strumous patient is a dangerous practice; but the injury inflicted on the recto-vesical floor admits of complete repair after a retentive catheter has been twice maintained in the bladder of the same patient for three weeks.

From a careful consideration of both cases I conclude that there is no evidence militating against a repetition of this operation as a *dernier ressort* in these miserable afflictions, and that it is our duty to lay such experimental facts before the surgical branch of our profession.

SUBJECT IX.

ABSTRACT OF A CLINICAL LECTURE ON THE TREATMENT OF STRICTURE OF THE MALE URETHRA.

DELIVERED MARCH 23, 1878.

('*Medical Examiner,' June* 20, 1878.)

GENTLEMEN,—I cannot direct your attention to a more practical subject than 'Stricture and its Treatment;' for unfortunately it is too common amongst men, and fortunately its surgical treatment is intelligible, facile, and restorative. Stricture is essentially a mechanical narrowing of the urethra, occurring generally at the membrano-bulbous tract; the sequence either of direct violence (traumatic); the result of nervous irritation (spasmodic); the outcome of gonorrhœa, aggravated by potent injections (gonorrhœal). When strictures are situate at the orifice of the urethra the more common varieties are cicatricial or congenital. But, without detailing varieties of stricture, or entering upon pathological statements, let me at once suppose that the commonest form of stricture, viz., gonorrhœal, is under our surgical care; then what treatment is a proper one to overcome this mechanical obstruction, and is in our judgment the best for the patient? In this hospital, where Mr. Barnard Holt has for so many years publicly taught and practised *splitting,* as the immediate remedy for stricture, one naturally gives this method the preference. I have had the good fortune not only to watch and study Mr. Holt's cases in our male wards, but also to have treated forty cases of stricture personally—twenty-two by the use of his own instrument, and eighteen by the use of a modification of Holt's dilator.

And here let me offer a remark to surgeons with reference to the superiority or inferiority of particular methods in surgical practice.   My association with Mr. Holt as a colleague has, undoubtedly, conduced towards training me in the use of dilatation, and has convinced and confirmed me in my acceptance of his doctrines; in the same way as many a Conformist or Nonconformist owes his actual creed to the accidents of his nativity and subsequent associations.   Apart from my early predilection, however, conviction on the value of this treatment has followed on analysing the success of dilatation, the freedom from complication, the absence of free hæmorrhage, the rapid recoveries, and the immunity from recurrence under wise precautions.   I say unhesitatingly that when you are surgeons, if you know as a fact that a certain procedure yields definite results, and is unattended with danger, you need not waste your time by experimenting on the comparative value of cutting or splitting urethral strictures, but you may fairly anticipate similar good results by the application of similar means.   Use the instruments that you have become familiar with—it is my fault if you are entrusted with bad ones—and, having made your selection, perfect yourselves in handling them.   King David did more execution with his sling than he felt able to do with Saul's elaborate armour.   Nor be too eager to level criticism at the heads of those who practise towards the same end by other methods.   In the same way as many roads converge towards London, and many locomotive methods exist for reaching it, so also varying treatment may successfully lead to the alleviation of stricture; but in my opinion it would be equally unsafe to ask an engine-driver to handle the ribbons, or a cabman to pilot an engine, as it would be to ask a surgeon to skilfully direct unaccustomed instruments.   A surgeon does more good by actually stating his own facts and his own results than by pointed invective against men better versed than himself in the very surgery that he denounces.   For example, I use immediate dilatation for strictures, because facts and results justify this practice; and my experience has an argumentative value; but I am not justified in condemning, wholesale, surgeons who perform internal urethrotomy, for I have never performed the operation, nor ever even watched clinically an example of it.

During the past decade 115 cases of urethral stricture have been split in the wards of the Westminster Hospital, with three deaths, showing a mortality of 2·06 per cent.   One of these occurred in Mr. Holt's practice, one in Mr. F. Mason's, and one in my own.

I will now exhibit to you the dilator (figs. 19), modified by me for general use in cases of stricture:—

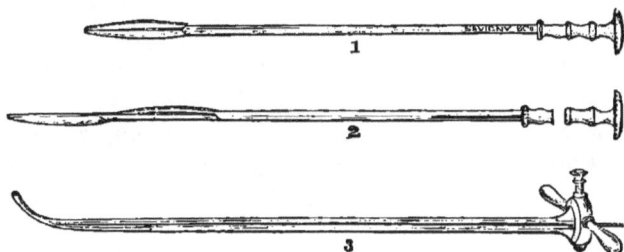

FIGS. 19.

Fig. 1.—The bolt set up.

Fig. 2.—The bolt divided into three parts, viz., two sliding rods, with bulbous-shaped ends; and the round handle, which screws on to the end of one rod, and fixes both, as seen in fig. 1.

Fig. 3.—The dilator, with central tube from tip to handle, for the transit of urine or catgut.

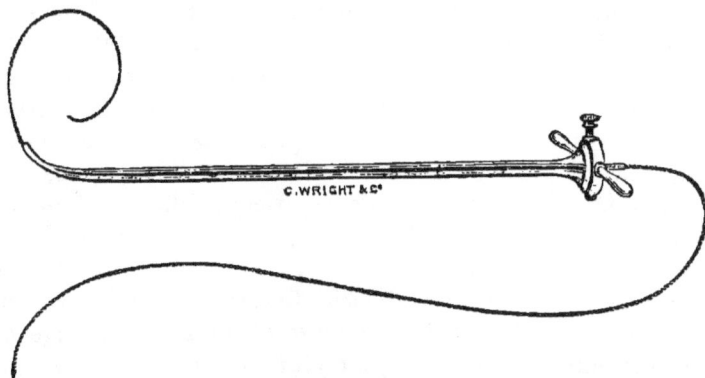

FIG. 20.

Fig. 20.—The curved line shows the catgut traversing the central tube.

In operating carefully introduce the point of the dilator down to the stricture, and try if the dilator will traverse it; if it does pass into the bladder, so much the better; slide in your bolts one by one, unite them, and push home. But if you meet with obstruction, and are unable to pass the dilator, maintain the tip of the dilator on the stricture, and manipulate the catgut conductor, and seldom will you fail to pass this *avant-courière* through the stricture. Slide the dilator on into the bladder, using the catgut as a single-line railway.

With reference to your being positive (before splitting) that the dilator is in the bladder, you cannot err with this modified dilator; if the bladder be partly full, urine follows the withdrawal of the catgut; if the bladder be empty, you can always slide on inch after inch of the catgut into an empty organ; for this catgut (limp as it is after usage) proves that the tip of the dilator is in an empty space. Now, the only three possible empty spaces in the male pelvis are the bladder, the rectum, and the peritoneal cavity. An index-finger in the rectum at once settles the presence or absence of the catgut in the rectum territory; while the undertaker will most likely clear up any doubt, in the event of the almost impossible introduction of the catgut into the human peritoneal sac. This accident, however, has happened in a case of ascites. On withdrawing the dilator unscrew the handle of the bolt, and remove it *sin gulatim.* You will always have some hæmorrhage from the urethra, and this must come from the torn vessels of the mucous membrane. In the two cases I have examined after death the mucous membrane was linearly torn in both; and in every instance (five) in which I have split strictures of the urethra on the post mortem table the subsequent dissection has demonstrated a ruptured mucous membrane in one or more places. I account for Mr. Barnard Holt's experience differing from my own by reason of many days having intervened between the date of operation and date of death, thus permitting an absence of scar by the rapidity with which injuries to the mucous membrane of the urethra unite. I know of no mechanical agency which so surely spares healthy structure, and so unfailingly challenges the strictured parts, as the bolt driven home, between the blades, and on the conducting rod of the dilator. Immediately after the operation pass a No. 10 or 12 silver catheter, draw off the urine, place the man in bed, keep him warm, and give quinine and opium if necessary. The catheter that I am in the habit of using, and the bristle-brush for cleaning the cylinder, are drawn in the annexed figure (figs. 21). All the instruments are manufactured by Wright & Co., of 108 New Bond Street, W.

In conclusion let me say a word or two on the subject of gradual dilatation, as compared with splitting. This gradual dilatation

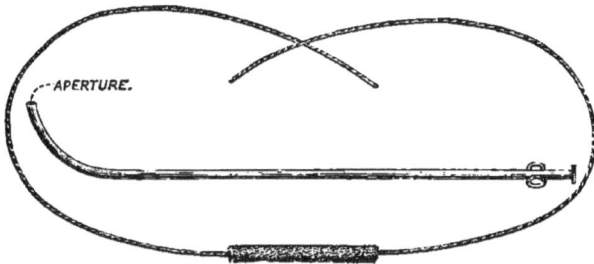

FIGS. 21.

might be styled *ad captandum* 'the gentle treatment of strictured urethra.' But when we consider that the majority of cases of traumatic stricture are due to the passing of catheters, and that false passages are not unknown quantities, for myself I would that the surgeon, armed with his uriniferous tubules, should make his visits, like those of angels, few and far between, than subject my person daily to the exercise of his gentleness. A neglected stricture is a present, past, and future enemy, the treatment of which should be, as in the engagements of the present day, sharp, short, and decisive; it brooks no delay nor gentle *de die in diem* adjournment. The time argument is a valuable one against gradual dilatation. How can a young shipmate the subject of stricture, with probably ten days' leave only, submit to this tardy improvement? Or, by way of illustration, supposing a dozen men, each thirty miles distant from their desired home, how many out of the twelve would travel by quick train or in a 'growler'?

The after-treatment of a dilated stricture is that of maintaining urethral calibre by the occasional (say once a month) introduction of a full-sized catheter. This precaution appertains to whatever treatment may be selected. It is undoubtedly a bother, and must be classed amongst personal inconveniences, but should be no more neglected than any other rule for the maintenance of health. You, as surgeons, may either instruct your patient to pass his own catheter or earn an honest guinea by introducing No. 10 for him. By so doing you maintain your dexterity, prevent the falling in of the tunnel, and uphold the success of one of the best operations in surgery.

SUBJECT X.

ABSTRACT OF A LECTURE ON THE SURGICAL ASPECT OF OUR PRESENT MODE OF RAILWAY TRAVELLING.

DELIVERED AT WESTMINSTER HOSPITAL, FEBRUARY 21, 1879.

(*'British Medical Journal,' March* 22, 1879.)

GENTLEMEN,—To some of you present to-day a short apology may be necessary for directing your attention to a subject which may seem to be without the pale of clinical surgery ; yet I have specially chosen the surgical aspect of our present mode of railway travelling as my text, because the prevention of grievous accident and the increased comforts of everyone are neither useless nor unattractive considerations. Moreover, I deem it to be the duty of all educated men, each in his own sphere, to direct public criticism to the persistent inattention of the railway officials of Great Britain to the safety and convenience of travellers. For myself I decline to accept the limited duty of doing the best that I can for an injured man, but consider it my privilege to avert the occurrence of harm, and to develop the truth of preventive surgery. In the same way that our physicians point out the blessings of sanitation and hygiene, and the means for excluding pests under the title of '*preventive medicine*,' so I maintain, it is more the duty of surgeons to prevent an accident than to prove their skill by aiding Nature in curing the recipient.

I cannot understand why the International Commission of Medical Men for the Limitation of the Plague (1878) has its *raison d'être*, unless you grant an equal standing to surgeons for the anticipation of preventable accidents.

You are aware that those who take upon themselves the office of critics are expected to suggest a remedy for their complaints; I will, therefore, connect with each specific charge a specific reform.

The habitual persistence of running human beings on the same line of metals on our main highways as goods and cattle calls for unqualified censure, for it is difficult to imagine that any mechanical arrangement could favour more collisions or produce more broken bones than the present. The system of mixing up goods and passengers on the same track may be compared to the absence of all passenger footways in a crowded London thoroughfare, with the advantage, however, that pedestrians yet might save their lives by their own activity; whereas, the rail on which passengers are travelling inevitably coincides with the rail on which are placed the heavy goods. To prevent this intolerable jumble, passenger-trains must be allowed their own special up and down line; and the safety of this arrangement is, to my mind, conclusively proven by the freedom from accident enjoyed by the Metropolitan Railway of London, over which but few carriages are run excepting those literally teeming with human beings.

Let us now watch the compulsory conduct of any traveller from a London terminus. The present system of issuing tickets at railway stations has conduced materially towards the production and aggravation of cardiac diseases. A definite number of persons have to be at a definite spot at a definite point of time. Most have experienced a mental anxiety not to be late for a train, added to an increased heart's action in hastening to catch it. His arrival is forestalled by many pressing towards a microscopical pigeon-hole; the inexplicit applicants for tickets or the dilatory clerk aggravate his want of patience, until, at last, his neck is dislocated in gaining a ticket and a bird's-eye view of the clerk. To prevent this unavoidable confusion railway authorities must encourage (by the opening of new offices and the ever-open office at the station) the practice of travellers providing themselves with tickets long prior to the departure of the train; or, in plain language, make the granting of a ticket at the starting-point the exception rather than the rule. Young women might as well elsewhere issue a pass over an open counter, as eclipsed clerks from their prison-cells at the station-house. Having taken his ticket, and personally seen his luggage labelled (by the way, what a grand institution is the American system of brass checks!), our traveller has to cross that dangerous chasm intervening between the platform and his carriage. This deadly ditch has much to answer for; and so repeatedly has this mechanical error been exposed, that any loss of life due to a passenger falling between the platform and wheels becomes absolutely manslaughter. I need scarcely say that the simple continuous footboard and uniformity of platform build prevent any further catastrophe under this count.

Next follows the tyrannical locking of the door or doors. I most strongly enter my protest against this autocratic barbarism. Cannot some British travellers be accredited with common sense? Will not the universal law of self-preservation insure others against harm? What right have railway directors to prevent a sane person from endeavouring to save his own life when in danger? Why should the majority of sensible travellers contract for imprisonment, and submit to it, in order that security may be granted to a drunken fool or an idiot? My blood boils at being (coolly) informed that a railway company insists on this dangerous confinement in order to economise labour in collecting tickets, as well as to demonstrate their anxiety to become the patrons of your safety. The quick satire of Sydney Smith is almost forgotten: * he advocated Episcopal martyrdom to bring directors to their senses. Surely the life of a director might well be paid in full discharge of this pretentious claim. Our recent disaster in Zululand, by shock, may contribute largely to our future success in arms; so the public would submit to the shock of an *Echo*-boy shouting, '*Special! Great Railway Disaster! One Director locked in and smothered!*' if this misfortune gained freedom for all. I have no ill feeling towards any director personally, and would be content if the loss of his arm or leg, or less, would suffice to arouse public attention to this intrusion on the liberty of the subject; yet, in order to gain emancipation, *if* one director *must* be told off, why, the sooner the better, for it might avert a repetition of the Abergele catastrophe, and, in the present temper of the times, I feel sure that one director would be enough. The ingenious handle now in use on the Metropolitan Railway is of itself a sufficient reform; and, by its general adoption, we need neither

---

* Letter to the *Morning Chronicle*, 1842.

fear self-destruction nor further irritation on the door-locking question.

Let us next inquire into the water-closet and urinal accommodation provided on British railways. They are all stationary, as if arranged for the further disquietude of the traveller, viz., to exact one more duty that must be performed prior to starting. The glittering buffet of Spiers & Pond, mark you, also waylays the starter who, thoughtlessly or by persuasion, has a glass of beer—by no means a gentle diuretic. Then, with his heart's action excited and kidneys stimulated, our traveller's corporeal torment commences. Whatever distance he may have to cover, the directors exact a visit to these stationary latrines (at the risk of losing the train) in order that Nature may relieve herself. Now, think for a moment, in disgust, at the practical alternatives presented to British passengers in the year of our Lord 1879. A rich man, probably in self-defence, purchases and submits to the repulsiveness of one of those putrid urine-bags known as 'railway conveniences'; while the bulk of suffering mankind are fain to let down the window and sprinkle indiscriminately their pent-up excretion. This is no joking matter, as you yourselves may have to admit its seriousness ; and, from surgical experience, I know that no surer way exists for a man to damage his kidneys and bladder, and the tone of his intestine and its gateway, than by an enforced restraint on the discharge of urine or fæces; and so imperative are Nature's commands on this point, that at last, after acute and intolerable agony, his pure ability of self-control is lost, and complete discomfiture reigns supreme. The American system has each car fitted with genuine railway conveniences ; and the adoption of a similar method in Great Britain is the direct solution of this grievous nuisance.

I shall next direct your thought to the gross absurdity of the present arrangements for communication between engine-driver, guard, and passengers. Who, think you, will smash his fist through an exaggerated watch-glass, and correctly turn a handle in the course of a struggle, or in a state of collapse? Or who will have the presence of mind (in anticipation of an assault) to ascertain on which side of an open window he or she may pull an inconveniently hung cord? Or, even assuming that such communication has been properly worked, of what service is it to know that in a few *minutes* (thanks to the non-introduction of continuous brakes) the train may be stopped, and the zealous guard making inquiries, after irreparable mischief has occurred ?   Or what avail or satisfaction can there be to anyone in the hands of a villain to imagine that his distress is witnessed by a third person, peeping through a glass, but unable to help by reason of an impassable partition ?   The days of our insular love of seclusion and approval of shut-up boxes must soon surely pass away, and lead to the public insisting that through communication in trains is the safest, best, and most civilised plan of carriage construction.   I have watched personally the comfort and safety of American cars, and affirm that indecent assaults and impudent proposals are the sequence of our mechanical build of carriages, and that such occurrences are to be rendered impossible by the adoption of the through thoroughfare in carriage construction.   The office of a guard to a railway train is analogous to that of the hands in our own economy : Nature has decided that our hands shall be able to immediately approach and succour any part of our bodies externally—of what immediate use are active hands, with paralysed arms and forearms, as compared with the utility of sound perfection ?

Lastly, let me insist on the necessity for reform in providing suitable sleeping accommodation for weary souls and bodies during night travel.   I am not here pleading the interest of those able to enjoy the luxury and expense of a Pullman's car, but the comfort and time-economisation of professional, commercial, and working-men.   I think that my own individual experience warrants me in stating as a fact that the use of hammocks on railways affords a most enjoyable and inexpensive means of securing rest during night-travel.   I have slept on our British railways thousands of miles in a hammock, and quite shudder when the alternative of a closely confined and penned-up attitude on a rectangular seat is presented to me in lieu of the mental and bodily repose imparted by the use of a portable swing-bed.

I would just mention how time is wasted by the present system of stopping to collect tickets.   For instance, in every journey between Waterloo and Exeter two special halts take place : one at Salisbury, the second at Exeter.   The through communication in carriages obviates this daily loss of time, and eliminates one source of danger-begetting unpunctuality in train-conduct.

Let me plead also for the remission of the abuse of the steam-whistle at starting and at all times. Any musical explosive short note is as quickly interpreted as a prolonged scream. One of the problems of the present day is to avert noise and screen from annoyance our auditory nerves, the most sensitive and delicately hung brace of strings; and yet even now in London, in addition to the inevitable reverberation of traffic and hum of business, the patient brain-workers have to endure the Italian organ-grinder, the muffin-bell, the mendacious shouts of newsmongers, and the livelong day and night penetrating shrieks of the engine host.

I have entered these protests and suggestions as the outcome of practical observation, and in order that the expenditure of money may not be allowed an undue argumentative value in the question of preventing actual loss of life or accidents, and in the firm belief that it is our duty as surgeons to increase the comfort and wellbeing of our fellow-men. My hope is that soon the increasing necessity of travel will compel the Government to relieve multifarious Boards of their multifarious duties; and that the conduct and control of British railways will be allied with the Post Office and Telegraph as State Departments.

At the present time, although the train speed and service is moderately good only, our railway directors' conduct in matters of civilised locomotive detail is extremely bad, and, as usual, in fertility of resource, . inventive adaptation, and utilitarian progress they have been fairly outstripped by our American cousins.

SUBJECT XI.

LECTURES ON AMPUTATION AT THE HIP-JOINT,
ILLUSTRATING THE USE OF THE RECTAL LEVER.

By RICHARD DAVY, M.B., F.R.C.S.,
Surgeon to the Westminster Hospital.

PREFACE.

' Have by some surgeon, Shylock, on your charge,
To stop his wounds, lest he do bleed to death.'
PORTIA: *Merchant of Venice.*

HAVING shown that the *per rectum* compression of the common iliac artery has many important surgical aspects, and my own experience having been confirmed by other operating surgeons, I have summarised our recorded observations.

Adducing the blue line on the gums in lead-poisoning as a clinical observation by itself worthy of a lifetime, I trust that the introduction of another fact into surgical practice may prove of interest and utility.

Knowing, moreover, that in general dealing it is but fair to submit the mouths of our favourite hobbies to judicial inspection, I have appended the published reports of other surgeons, who are more competent than myself to pass an impartial judgment.

LECTURE I.

CLINICAL LECTURE ON AMPUTATION AT THE HIP-JOINT.

CASE ILLUSTRATING A NEW METHOD OF COMPRESSING THE COMMON ILIAC ARTERY.

DELIVERED AT THE WESTMINSTER HOSPITAL, FEBRUARY 13, 1878.

(' *British Medical Journal,*' *May* 18, 1878.)

GENTLEMEN,—I shall endeavour to-day to engage your attention on the gravest amputation in surgery, viz., that at a hip-joint. For the practical elucidation of this Lecture I will read the notes of the boy's case who is now in this theatre; next I will describe in detail the method of operating; and will, lastly, carry out the whole procedure on the dead body.

C. C., aged 9, schoolboy, was admitted into Mark Ward on July 10, 1876, suffering from morbus coxæ (femoral and acetabular) on the right side. In 1874 he had two severe falls, one especially from a cliff near Yarmouth. Pain in the right hip and lameness set in at once. Some months afterwards swelling and puffiness were noticed over the right hip; the abscess, however, did not open over the articulation, but burrowed along the adductor group of muscles; and, in June 1875, a sinus formed at the junction of the middle and lower third of the thigh. Suppuration was watery and profuse; the limb became everted and drawn up towards the abdomen. He was admitted because alarm had arisen from his emaciation. From July 1876 until January 1877 he was kept in the recumbent posture; had a most generous diet, and quinine; rest to the joint was maintained; and at times he was slung in the open air in a hammock. The boy was so surely losing ground that, on consultation, it was agreed to remove the right thigh at the hip-joint.

Method of operating.- In all severe amputations one of the first considerations of the surgeon is to anticipate shock and to prevent the loss of blood. I am inclined to permit any patient to have a glass of wine or brandy-and-water about one hour before the operation: the result partakes more of a sedative than stimulating character, apprehension is lessened, cardiac tone is gained, and fitness for the ordeal is exhibited. The American surgeons, with their characteristic ingenuity, devised pressure on the abdominal aorta for hæmostatic ends during amputations high up towards the pelvis. Lister arranged a horseshoe clamp and screw-pad for compressing the aorta above the umbilicus. I saw this mechanism (1860) used in Syme's brilliant operation on gluteal aneurism at the Royal Infirmary, Edinburgh; but in 1874, during the delivery of the Surgical Lectures at the College of Surgeons, in London, by Professor Holmes, I drew the lecturer's attention to the possibility of controlling the aorta, common iliacs, and internal iliacs by pressure through the rectal wall, and wrote the following note to the ' British Medical Journal ':—

' June 20, 1874.—*Compression of the Internal Iliac Artery.*—Sir,—Mr. Holmes, in his lecture on Gluteal Aneurism, at the Royal College of Surgeons, June 8, mentioned the late Mr. Syme's case, treated by boldly laying open the sac, and restraining the hæmorrhage by Lister's aortic clamp. It may be well to draw surgical attention to the fact that the internal iliac artery may be effectually compressed for a time by pressure through the rectum on the true pelvic wall; a less serious procedure than compression of the aorta through the abdominal wall.—I am, &c.,   RICHARD DAVY.'

On January 16, 1877, this case presented a fair opportunity for testing the suggestion. Having passed his urine, the right leg and thigh were emptied partially of blood by Esmarch's bandage. Chloroform was administered, and about one fluid ounce of sweet-oil was sent up his empty rectum. A straight lever of wood (run smooth and round out of a lathe) was introduced *per rectum;* the small end was applied over the right common iliac artery between the lumbar bodies and psoas magnus muscle; the projecting part of the lever ran nearly parallel to the left thigh.

My colleague Mr. Thomas Bond readily compressed the common iliac artery by elevating the projecting arm of the lever, the perineal tissues acting as a fulcrum. Accordingly, as Mr. Bond elevated or depressed the lever so did the right femoral artery cease or continue to pulsate. The left femoral

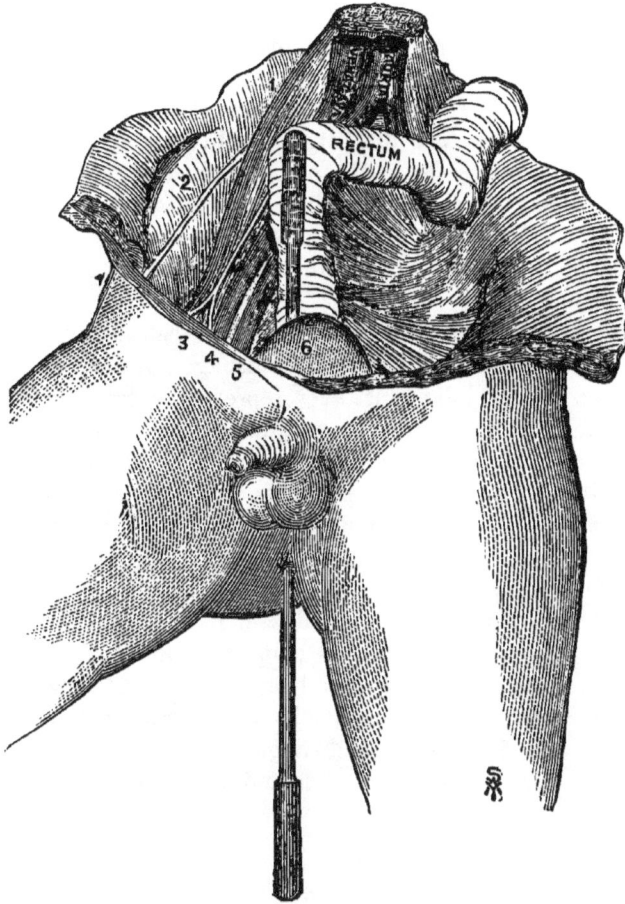

FIG. 22.—The rectum is supposed to be transparent. 1. Psoas magnus muscle. 2. Iliacus muscle and external cutaneous nerve. 3. Genito-crural nerve. 4. External iliac artery. 5. External iliac vein. 6. Bladder.

was undisturbed, beating with regularity throughout. A long square anterior flap was made by transfixion over the joint; the muscles and capsule were divided, and a short posterior cut severed the limb. The arteries were tied, sutures inserted, and the boy placed in bed.   About a wineglassful of blood only was lost.

January 30.—The stitches were removed.

February 3.—The ligatures were removed.

May 16.—He was discharged convalescent.   Three sinuses were together weeping half an ounce of pus *per diem.*   The boy looked fresh and well.

February 15, 1878.—The boy says now that he is quite well.   Two sinuses are discharging a little, and that little is becoming less every day.   He gains weight and flesh.

My finger in the acetabulum readily entered his pelvis during the operation; a loose piece of bone was removed; and this pelvic mischief accounts, in my opinion, for the yet present sinuses.

In this operation I have demonstrated that surgeons possess a new method of controlling hæmorrhage; and, as an amputation at the hip-joint must ever represent a grievous loss of tissue, any amelioration in its performance must be welcome to the surgeon and the patient.   I venture to say that compression of the iliac artery *per rectum* is less serious than compression of the aorta through the abdominal wall; that the former is more *easy and reliable* than the latter, a straight smooth lever alone being required; that the circulatory system is far less disturbed, by reason of the circulation in the opposite limb remaining undisturbed; and no danger results to the rectum from the lever, if guided by the hand of an expert surgeon.

[Mr. Richard Davy concluded by demonstrating the operative details on the dead body; and suggested that this new method of compressing the aorta and iliacs might apply to many operations in pelvic surgery.]

LECTURE II.

CLINICAL LECTURE ON AMPUTATION AT THE HIP-JOINT.

WITH CASES ILLUSTRATING THE USE OF THE LEVER FOR CONTROLLING HÆMORRHAGE.

DELIVERED AT THE WESTMINSTER HOSPITAL, AUGUST 22, 1879.

(*'British Medical Journal,'* *November* 1, 1879.)

GENTLEMEN.—Some few years ago I drew the attention of surgeons to the *per rectum* compression of the aorta and iliac arteries for the temporary arrest of hæmorrhage during the performance of operations, and practically demonstrated the efficacy of the method in cases of amputation at the hip-joint. I shall endeavour to interest you to-day, first, by practical observations on amputation at the hip-joint; and, second, by recording the ascertained experience of operating surgeons in the use of the lever for controlling bleeding during amputation at the hip-joint.

Firstly, in considering how surgeons may best restrain bleeding in amputation at the hip-joint, just look at a side-view of the pelvis, with the arteries injected (Royal College of Surgeons' Museum, 939, Q. f. Homo); you will notice how that the anterior flap is principally fed by the external, the posterior by the internal iliac artery; therefore, the common iliac artery must be compressed to secure both. And Nature has so focussed her blood-courses at the common iliac artery for the nourishment of the lower limb, that it really seems strange that compression of this vessel should not have been earlier practised.

I would insist strongly on the absolute necessity for restraining loss of blood from the posterior flap. Much as I respect the femoral artery in the anterior flap, yet more do I beg you not to underrate the large and numerous muscular branches of the gluteal and sciatic trunks in the posterior. In an interesting discussion at the Clinical Society of London, following a paper read by my colleague Mr. A. Pearce Gould (*vide* 'British Medical Journal,' May 10, 1879), the use of the rectal lever was received with favour, and a comparison was drawn between it and digital compression. I can only state that, after a surgeon has proved practically how very easy it is to prolong indefinitely compression by the lever, he will not care to submit to the fatigue, uncertainty, and anxiety of the digital pressure. Nor can I see why the whole thickness of the abdomen should be squeezed, now that it has been shown that the common iliac artery may be compressed, with only the thin rectal wall intervening.

Mr. Newman of Glasgow has sent me a pamphlet in which he describes a special knife and india-rubber bands for compressing the flaps during amputation at the hip-joint. It seems to me that, as Pancoast's tourniquet applied over the aorta is too high up, so also Mr. Newman's apparatus on the flaps is applied too low down; and that operating surgeons in future will act on the spirit of the motto,

'In medio tutissimus ibis,'

and compress the common iliac artery *per rectum.*

I will just say one word on the perfect safety and absence of inconvenience in using the lever. I have personally compressed the common iliac artery in a man, aged 60, suffering from aneurism of the right external iliac artery, for twenty minutes, and without chloroform or other anæsthetic. The only pain felt was the presence of a foreign body, showing how tolerant of pressure the upper part of the rectum is. In no case (so far as I am aware) has any blood or stain been seen in the stool, though I have watched carefully for this event.

On anatomical grounds the situation of the common iliac artery is perfect for the ends of compression; the lever drops between the psoas magnus muscle and the bodies of the lumbar vertebræ, having the spring (sacral margin) of the true pelvic brim as a counter-resistance to the lever, and no large nervous trunks in the way. I tried to pass the lever once on a dead body at the Westminster Hospital, and failed. I dissected this rectum *in situ,* and found syphilitic stricture of it; in short, the gut was tied up like a boy's cracker. This is the only case in which any difficulty has been met with at present.

The lever is turned out of ebony, and varies in length from

eighteen to twenty-two inches. Its surface is very smooth and polished, and its ends are rounded off much like the finger-tips. The maximum transverse diameter is five-eighths of an inch; the minimum, three-eighths of an inch. The rectal end is graduated to an inch scale, so that the surgeon who applies the

C WRIGHT & C°

Fig. 23.

lever can at once learn whereabouts may be the end of the rod.*

I will end my first heading by enumerating what I maintain to be the salient advantages of rectal compression :—

1. Most perfect control of the required artery.

2. Minimum amount of disturbance of the circulatory system.

3. Independence of the respiratory movements, of great importance from the chloroformist's point of view.

4. Its general and easy applicability; strictured rectum being the sole obstacle.

5. The pressure applied is so easy to maintain, and the assistant's body so well out of range of the operator, that no hurry need perplex the one, nor anxiety the other.

6. Its application is quite safe in skilled hands, no injury having ever resulted, and but little pain having been suffered.

7. Cheapness and simplicity, illustrating a lever of the first order.

8. The success hitherto achieved by its employment.

Let us now consider the already ascertained experience of operating surgeons on the use of the lever in controlling bleeding during amputation at the hip-joint.

I have the record of ten cases in which the lever has been used. The total amount of blood lost during the ten operations has been under seventeen ounces, and there have been 80 per cent. of recoveries.

Here is a table, drawn up under my own observation, which summarises this new experience. The cases have occurred in the practice of Mr. Gould, Mr. Armstrong, Mr. Young, Mr. B. Wills Richardson, Mr. Cadge, Mr. George Cowell, Mr. Macnamara, Mr. Stokes, and myself.

* The instrument is manufactured by Messrs. Wright & Co., of 108 New Bond Street, W.

G

Table of Cases of Amputation at the Hip-joint.

No.	Sex	Age	Admission	Year	Surgeon	Nature of Disease	Operation	Departure	Result	Ounces of blood lost
1	Boy	9	July 10	1876	Mr. Davy .	Morbus coxæ. R.	Jan. 16, 1877	May 16. .	Recovery	1½
2	Man	28	Oct. 26	1878	Mr. Bond / Mr. Gould	Morbus coxæ. R. Excision previously	Dec. 7, 1878	December 11	Death from thrombosis	3
3	Boy	12	Feb. 2	1878	Mr. Davy / Mr. Armstrong	Morbus coxæ. R. Excision previously, Dec. 17, 1878	Jan. 22, 1879	January 31 .	Death from exhaustion; albuminuria	2½
4	Boy	9	Jan. 12	1879	Mr. Young . / Mr. H. Marsh	Morbus coxæ. R.	Feb. 4, 1879	May 29. .	Recovery	1½
5	Girl	8	Feb. 14	1879	Mr. B. Wills Richardson (Dublin)	Morbus coxæ. L.	March 1, 1879	June 10. .	Recovery	2½
6	Man	21	April 12	1879	Mr. Cadge (Norwich)	Necrosis, acetabular and femoral Osteo-sarcoma. R. Fracture of lower third of femur	April 4, 1879	June 9 . .	Recovery	2
7	Boy	7	Nov. 29	1878	Mr. Cowell .	Morbus coxæ. R.	June 17, 1879	September 3	Recovery	¾
8	Boy	5	Aug. 14	1878	Mr. Davy .	Morbus coxæ. R. Excision previously	July 16, 1879	In hospital .	Recovery	1
9	Boy	6	May 7	1879	Mr. Macnamara . / Mr. Davy	Morbus coxæ. R.	Aug. 8, 1879	Eastbourne .	Recovery	1
10	Man	43	Aug. 29	1879	Mr. Stokes(Dublin)	Caries. Left .	Oct. 1, 1879	In Richmond Surgical Hospital	Recovery	1
									Total . .	16¾

Case 1.—This has been duly reported in my preceding Lecture.

Case 2.—Mr. A. Pearce Gould. This case is reported in the 'British Medical Journal,' May 10, 1879, page 704, and appears in the 'Transactions' of the Clinical Society. (See page 90.)

Case 3.—Mr. Armstrong, surgeon to the Royal Hospital School, Greenwich, has published his case in detail. (See page 101.)

Case 4.—Mr. Young of Sevenoaks also intends publishing his case in full.

Case 5.—Mr. B. Wills Richardson has reported his case in the 'Medical Press and Circular,' April 9, 1879. (See page 106.)

Case 6.—Mr. Cadge of Norwich wrote to me on June 9, 1879, as follows:—'I can scarcely think that any other plan than your lever will be used in future; I can see no drawback to its use, and the facility of its adaptability is equal to the certainty of its action, and in both these respects nothing better can be desired.' Mr. Cadge's patient died on June 15, 1879, of pulmonary sarcomatous deposits; but the stump had healed, with the exception of two sinuses.

Case 7.—Mr. George Cowell. This boy lost the smallest amount of blood ever recorded in an amputation of the hip; and the patient has made a rapid recovery.

Case 8.—This boy had lardaceous disease of his viscera, with distressing ascites. It would have been impossible to compress his aorta or his common iliac artery through the abdominal wall, by reason of the distension. This boy's recovery is pleasing, and the albumen in his urine is daily diminishing.

Case 9.—Mr. Macnamara. This boy made a splendid recovery, and was sent to Eastbourne one month after the operation.

Case 10.—Mr. Stokes of Dublin intends to publish his case *in extenso*. (See page 109.)

August 22, 1879.—I have this day received a note from Mr. W. Whitehead of Manchester, who states that he has used the lever with perfect success, and has forwarded his report thereon to the 'Journal.' (See page 108.)

It is worthy of remark that, out of these ten cases, nine are males, and eight occur in the right hip.

I was informed that Mr. Barwell had used the lever at

Charing Cross Hospital; his house surgeon, Mr. Pattison, has furnished me with the following facts:—

'On May 8, 1879, amputation at the left hip-joint was performed by Mr. Barwell at the Charing Cross Hospital.

'Boy, aged 7, the subject of morbus coxæ and previously excised left hip-joint. Mr. Davy's lever was used, and was held by Mr. Cantlie. The child lost a quantity of blood during the operation; this being seemingly due to the fact that the lever compressed the common iliac vein as well as the artery, the blood lost being principally venous. After the operation the boy recovered consciousness; but in about four hours' time was seized with a convulsion and died.'

My only remark on this case is, that it must be impossible for the branches of the common iliac vein to bleed had the common iliac artery been effectually compressed.

In order that I might verify the hæmostatic value of the lever in amputations at the hip-joint I performed the following experiments on the dead body:—Having opened the aortic sheath, I inserted into the aorta a large pipe about one inch and a half above the bifurcation into the two common iliac arteries, and connected this pipe with a water-syringe, compressed by the hand of our Surgical Registrar, Mr. Forsbrook. I next amputated the right thigh at the hip-joint, and the water ran out freely in a *per saltum* spring. I then inserted the lever *per rectum*, compressed the right common iliac artery (Mr. Forsbrook still continuing his pumping), and immediately the flow of water ceased at all the cut ends of the vessels, and remained so as long as the compression was maintained. The longest time that I remember in which the lever was acting is twenty minutes—more than sufficient time for a careful operator to complete the details of a hip-joint amputation.

The lever that was used for the first time on the living body, and which was applied on the common iliac artery by my colleague Mr. Thomas Bond, has been presented to the Museum (Surgical Instrument Department) of the Royal College of Surgeons, London.

In conclusion, I commend this simple method of restraining hæmorrhage to the notice of operating surgeons as safe, reliable and effectual. Its future is by no means limited to amputations at the hip, but is applicable to all surgical procedure where it is desirable to check *pro tempore* the blood-current through the aorta or iliac arteries; and I trust that the wide field of practical surgery may yield instances of its varied utility.

LECTURE III.

ON THE USE OF THE LEVER FOR AORTIC AND COMMON ILIAC COMPRESSION.

DELIVERED AT THE WESTMINSTER HOSPITAL, JUNE 11, 1880.

GENTLEMEN,—I propose to-day offering you some practical remarks on amputation at the hip-joint, and I will divide our subject-matter into two. 1. On the practice of ancient surgeons, and their results in amputation at the hip-joint. 2. On the present practice of surgeons, and their results.

1. This operation was advocated, in the first place, by French surgeons nearly a century and a half ago, for in 1743 M. Ravaton wished to perform it, but was prevented from so doing by his colleagues; and Perrault, in 1773, first operated successfully. The British surgeons, however anxious they may have been to play follow my leader, were influenced by the writings of Pott, who states, in the quaint language of professional trimming, 'that amputation in the joint of the hip is not an impracticable operation (although it be a dreadful one) I very well know. I cannot say that I have ever done it, but I have seen it done, and am now very sure I shall never do it, unless it be on a dead body.' The question of restraining hæmorrhage during the performance of this operation soon cropped up; for in 1803, we find Larrey advocating the prior ligature of the femoral artery; and at about the same time Abernethy taught that the femoral should be temporarily compressed against the os pubis. Sir Astley Cooper in his Lectures states, with reference to the compression of the femoral artery: 'I am disposed to think that the operation cannot be safely performed without securing the artery in the first instance. When you do not secure the artery in the first instance what is likely to happen is this—when you have to divide the femoral artery as near to Poupart's ligament as possible, and put a ligature upon it, the

man becomes so faint under the operation that he will be unable to support it.   I have in such a case been obliged to suspend it, to give the patient wine and chat with him, in order to rouse the vigour both of body and mind.   The operation will certainly be most safely performed by tying, in the first instance, the femoral artery, under Poupart's ligament, above the origin of the arteria profunda.'   In Sir Astley Cooper's operation it is stated that only about twelve ounces of blood were lost.

With reference to the formation and planning of flaps in this operation, Skey, in his 'Operative Surgery,' states, with grim humour : ' A pleasing variety of about eighteen operations for removal of the femur at the hip-joint have been practised.' I shall only advise you to exercise common sense, and apply it to each individual case ; keeping the skin intact where the scrotum rubs the inner side of the thigh ; saving as much soft tissue as you possibly can ; and having your front flap covering the wound epaulette-fashion, for the production of a dependent drainage, and the utmost simplicity of future dressing.   It is extremely difficult to accurately gauge the percentage of mortality after amputation at the hip-joint.   Undoubtedly traumatic cases and primary amputations are by far the most serious.   The highest rate of mortality reaches us from the Crimea, where 14 operations were performed, with 14 deaths.   The best table is that of Mr. Stephen Smith's, of New York, which gives a mortality rate of only 14 per cent. amongst 21 cases (1852).   Truth doubtless rests between the two.   I agree with Stokes in quoting the mortality rate after this amputation at about 50 to 60 per cent., and I am strengthened in this view by the calculation of all the operations quoted in Cooper's ' Surgical Dictionary.'   They amount to over 300 ; and the deaths are to the recoveries as 192 to 114, leaving a mortality of about 60 per cent.

Regarding these facts in history, one cannot but arrive at the conclusion that surgeons smote their patients hip and thigh with a great slaughter, as Samson did the Philistines.

We will now consider the second part of my Lecture, viz., the present practice of surgeons, and their results.   And here let me impress on you that the introduction of chloroform into surgical practice by that noble man the late Sir James Y.

Simpson marks the new era; and that salvation from pain is the prime pivot on which our success turns in these halcyon days.

Next, however, in importance for the safe conduct of an amputation at the hip-joint is the conservation of blood, especially if such an operation be delayed too long, as, in my opinion, it too frequently is. The statistical table of amputation at the hip-joint here given seems to prove conclusively that the mortality decreases *pari passu* with the amount of blood lost; or that, in other words, the salvation of blood implies the salvation of a patient (59 per cent. recoveries); and it is but reasonable to admit that in a debilitated case shock and great loss of blood are the last straws (and not light ones) that break the camel's back. I have used a modification of the rectal lever, as shown in the annexed sketch, and it has the advantage of permitting the surgeon to introduce A, the flexible bougie, into the rectum while the patient is under an anæsthetic in his own bed, and incurs no risk to the bowel during transit from the bed to the operating theatre. On commencing the operation the rigid rod B is introduced into A, and the lever is manipulated in the usual manner.

FIG. 24.—A, the elastic waxed bougie, capable of introduction in the curved form. B, the rigid lever.

The annexed table gives our surgical experience in the use of the lever; and I am much indebted to all the surgeons for their courtesy in furnishing the facts recorded.

In addition to many public demonstrations in our own theatre, I have had the honour to be asked to control the common iliac artery in amputation at the hip-joint in the theatre of my old school, Guy's (Mr. Bryant); as well as in the operating theatre of the Middlesex Hospital (Mr. Henry Morris). I have in my father's village surgery spilt more blood at one sitting in common venesection, than 17 patients have lost in 17 amputations at the hip-joint. A lusty countryman would not succumb under a quart of blood, while the whole amount of blood lost in these 17 amputations is only 36¼ fluid ounces. It is a pleasing reflection, as years roll on in surgical practice,

Table of Amputations at the Hip-joint. Use of Davy's Lever.

Age	Admission	Year	Surgeon	Nature of Disease	Operation	Departure	Result	Oz. of blood lost
9	July 10	1876	Mr. R. Davy	Morbus coxæ. R.	Jan. 16, 1877	May 16, 1877	Recovery	1½
28	Oct. 26	1878	Mr. T. Bond	Morbus coxæ. R. Previous excision	Dec. 7, 1878	Dec. 11, 1878	Death from thrombosis	3
12	Feb. 2	1878	Mr. Gould	Morbus coxæ. R. Previous excision, Dec. 17, 1878	Jan. 22, 1879	Jan. 31, 1879	Death from exhaustion. Albuminuria.	2½
9	Jan. 12	1879	Mr. Armstrong	Morbus coxæ. R.	Feb. 4, 1879	May 29, 1879	Recovery	1½
8	Feb. 14	1879	Mr. Davy, Mr. Young	Morbus coxæ. L. Acetab. and fem. necrosis	March 1, 1879	June 10, 1879	Recovery	2½
21	April 12	1879	Mr. Marsh	Osteo-sarcoma. R. Fracture of lower third of femur	April 4, 1879	June 9, 1879	Recovery	2
7	Nov. 29	1878	Mr. Richardson (Dublin)	Morbus coxæ. R.	June 17, 1879	Sept. 3, 1879	Recovery	¾
5	Aug. 14	1878	Mr. Cadge (Norwich)	Morbus coxæ. R. Excision previously	July 16, 1879	Oct. 21, 1879	Recovery	1
6	May 7	1879	Mr. Cowell	Morbus coxæ. R.	Aug. 8, 1879	Sept. 8, 1879	Recovery	1
43	Aug. 29	1879	Mr. Davy, Mr. Macnamara	Caries. L.	Oct. 1, 1879	Dec. 15, 1879	Recovery	1
26	Sept. 14	1879	Mr. Stokes (Dublin)	Morbus coxæ. Excision, Sept. 22, 1879	Oct. 15, 1879	Oct. 15, 1879	Death, shock	1
8	Nov. 28	1879	Mr. Whitehead (Manchester)	Caries. L. Dislocation. Sequel of scarlet fever	Jan. 20, 1880	June 24, 1880	Recovery	3
13	Nov. 17	1879	Mr. R. Davy, Mr. Bennett (Cheltenham)	Strumous disease of right hip. Excision	Jan. 2, 1880	Jan. 4, 1880	Death from exhaustion. Albuminuria	2½
21	Feb. 25	1880	Mr. Bryant, Mr. Davy	Osteo-sarcoma of right femur. Fracture	March 5, 1880	May 11, 1880	Recovery	3
17	Jan. 19	1880	Mr. Morris, Mr. Davy	Old morbus coxæ. Left pelvic necrosis	May 19, 1880	May 21, 1880	Death. Diseased viscera and peritonitis. Chronic	2
10	Aug. 1	1879	Mr. R. C. Lucas	Right morbus coxæ. Pelvic abscess. Excision	Nov. 15, 1879	Jan. 7, 1880	Death nine weeks after, from lardaceous disease	4
57	Jan. 29	1880	Mr. R. C. Lucas	Crushed left femur. Railway accident	Jan. 29, 1880	Jan. 29, 1880	Death one hour after. Shock	4

that one has been the means of saving some blood in atonement for what has been shed.

On June 25, 1880, my colleague Dr. Potter had occasion to remove an epithelioma from the uterus (cervix had been previously excised); on consultation we thought that the hæmorrhage might be lessened by a temporary compression of the aorta. The aortic lever was introduced *per rectum*, and compression was maintained for about three or four minutes; the femoral arteries both ceased to pulsate; the growth became livid, and the circulation was arrested *pro tempore* distal to the point of pressure. Dr. Potter intends to bring the case before the Obstetrical Society of London. In the meantime I must thank Dr. Potter for giving me the opportunity of performing for the first time on a living subject aortic compression *per rectum*. I had performed many experiments on the dead body, and had waited anxiously for the practical test of my observations on an actual case. In my belief the aortic lever will prove of further utility in controlling the uterine arteries (*e.g.* post partum hæmorrhage; excision of uterus); and its future value may safely be left to the estimation of skilled obstetric operators. I have drawn a copy of the steel aortic lever; it is suited to the pelvic

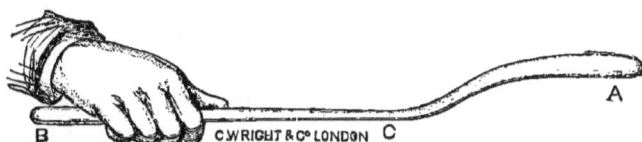

FIG. 25.—A, gouge-shaped end. B, handle of lever, which might be used for compressing the common iliac artery. C, point of fulcrum.

curve, and one end expands into a flat gouge-shaped surface, which on elevation of the handle clips the aorta against the convexity of the anterior bodies of the lumbar vertebræ. Some modification of the fulcrum will have to be worked out, for the elevation of the recto-vaginal wall materially tends to obstruct the vagina, and consequently interferes with the complete freedom of the operator's fingers.

In conclusion I may state that up to the present the good results gained by the use of the lever have been positively present, and ill effects negatively absent; it is a mechanism

requiring care and knowledge ; and, receiving such in the hands
of operators, I confidently assign to it a recognised place in
surgical practice.

CASE OF AMPUTATION AT THE HIP-JOINT IN WHICH THE COMMON ILIAC ARTERY WAS COMPRESSED PER RECTUM.

BY A. PEARCE GOULD, M.S. READ APRIL 18, 1879.

(*'Clinical Society's Transactions,'* vol. xii. p. 167.)

THE following case was under my care at the Westminster
Hospital, having been sent to me by Mr. Morris of Edmonton;
Mr. Macnamara kindly allowing me to use one of his beds.

W. G., æt. 28, a gardener, stated that in childhood and
again eight years ago he had an abscess in his right hip, which
healed up after discharging for some time. At Christmas,
1877, an abscess formed there again, and burst in two places
on the outer side, and discharged profusely and continuously
up to his admission on October 26, 1878, other sinuses having
formed in the meantime. He had become very anæmic,
emaciated, and depressed, but was not suffering from hectic.
On examination under ether, it was found that in addition to
advanced disease of the articulation, the neck of the femur was
fractured and gave rough crepitus on moving the limb. All
the sinuses were discharging thin ill-conditioned pus. There
was no induration or sign of disease in the pelvis or in the
other organs of the body.

October 30.—I excised the head of the femur through an
incision prolonged up from a sinus above the great trochanter.
I had some difficulty in getting the detached head out of the
acetabulum, but accomplished it by means of the lion forceps and
periosteum scraper. The entire surface of the acetabulum was
bare. The wound granulated, and at first all promised well;
the man became brighter, took his food better, and slept soundly.
But the old sinuses continued to discharge. I tried to remedy
this by securing free drainage, and on three successive occasions
I passed drainage tubes in various directions for this purpose.

In particular one sinus on the part of the inner side of the thigh continued to discharge thick laudable pus in large quantity in spite of a free dependent opening, and it was ultimately found that pressure over the psoas muscle above Poupart's ligament caused this pus to escape more quickly. At the same time the patient began to suffer from night-sweats. On December 4, the thigh became greatly swollen, tense, and œdematous, with dropsy of the foot and leg; the temperature rose to 102° and 103·6° at night; pulse 102, weaker. Two days later on the man was very sick; an abscess burst close to the sinus mentioned above, thigh still greatly swollen. Urine contained a trace of albumen. At a consultation with Mr. Macnamara, it was decided that amputation at the hip-joint offered the only chance of saving the patient's life.

Up to this time the general treatment had consisted in a liberal diet, containing a chop or slice of meat and two eggs every day, but no stimulants. Quinine was given at first, and after a few days sulphide of calcium in doses of $\frac{1}{8}$ grain every four hours was added, with a view of lessening the suppuration; in this it failed, and it was discontinued on the occurrence of diarrhœa. Then sulphuric acid was given alone as a tonic and antisudorific with apparent benefit. It is worthy of note that the patient's digestive organs were in a very good state, the tongue was clean and moist, his appetite was good, he enjoyed his food, and the bowels acted regularly, with the one exception named above.

At 10 A.M., December 7, he had a mutton-chop, and at noon half an ounce of brandy in some tea.

Operation.—At 2 P.M. he was put under the influence of ether. Esmarch's bandage was applied from the toes to the middle of the thigh, and the pubes, buttock, and groin were shaved and then well washed with carbolic lotion (1–20). Mr. Davy passed his 'lever' up the rectum for a distance of 9 inches, and by gently raising the handle at once stopped the pulsation in the femoral artery. The limb being held by Mr. Neale, I then amputated it at the hip-joint by the oval method, beginning the incision at the lower end of the excision wound, carrying the knife lightly across the front of the thigh, but down to the bone on the inner side and behind. After forming the posterior flap I sank my knife deeply in on the front and removed the limb. Having ligatured the femoral vessels and profunda artery with carbolised silk, and several other vessels with catgut, I washed the flaps and sinuses with carbolic lotion (1–20), and placing a strand of horsehair between the flaps I brought them together by means of three deep and several superficial sutures, and applied a common antiseptic dressing. The man was then removed to a warm bed. All the soft tissues of the thigh were immensely œdematous and infiltrated, the skin and cellular tissue being from an inch to an inch and a half in thickness; serous fluid poured from the parts during their division. I must acknowledge the very able help rendered me by Mr. Macnamara, especially in the arrest of the hæmorrhage from the very numerous smaller vessels.

The patient suffered considerably from shock, and was sick during the next five hours. At 10 P.M. I found him quiet and comfortable, recovering nicely from the collapse, with a warm and moist skin; he had taken some beef-tea, had just passed urine, and only a little serous discharge had oozed through above the dressing; no blood was to be seen. He dozed through the night, and in the morning was quite bright and cheerful, and said he had 'no pain to speak of.' Temperature 100°; pulse 120. Tongue moist and quite clean. The dressing was renewed; the one removed contained a relatively small quantity of bloodstained serum, no clots. The flaps were in accurate apposition—no tension. The urine passed contained excess of urates and a faint trace of albumen. The patient had taken a good quantity of beef-tea and asked to have a chop for dinner. This he had and enjoyed, and afterwards had some sound sleep; no pain. At 10 P.M., he was quite comfortable; no escape of discharge from the dressing. He slept soundly through the night, and woke at 5 A.M., December 9, faint and low, in a profuse sweat, which continued for some hours. At 10 A.M. his pulse was 120, tongue moist and quite clean. Dressing changed; discharge free and good; the lower posterior corner of flap was seen to be sloughing over an area the size of a crownpiece; the superficial suture at this corner was removed. No redness of flaps nor tension. Urine loaded with urates, and dark olive green in colour from carbolic acid; in the 24 hours ending at 12 noon he had passed 20 oz. Protective was laid over the wound, and lint soaked in carbolic oil over that. He took a little fish for dinner, and had an ounce of port wine in the afternoon. He was quiet but 'low' all day. In the evening the dressing was renewed and terebene lotion substituted for carbolic oil. He had a very restless night, complaining of pain in the left leg, and passed three loose motions—no mucus or blood in them. At 8 A.M., December 10, he was sick and then became notably worse—very distressed, with hurried respiration, severe pain in left leg and foot. His face was blanched, there was no cyanosis, air entered his chest freely, and he was quite conscious. The left femoral artery was felt pulsating forcibly; the superficial veins of the limb were very distinct; œdema of foot, but none of leg or thigh—the whole limb was warm. He gradually became unconscious, and died at 11 A.M., 69 hours after the operation. His tongue was clean and moist up to the time of his death.

Autopsy, 27 hours after death.—Lungs not congested, a little old tubercle at each apex. Heart contained fluid blood on the right side and in the pulmonary artery, and a firm fibrinous coagulum in the left ventricle and aorta. Alimentary canal, liver, and brain healthy. Spleen large; on section its dark surface was seen to be mottled towards the centre with lighter red patches. Kidneys pale and swollen. Rectum was carefully examined. No tearing or bruising of the meso-rectum; no sign of injury to the rectum visible from the outside, and on slitting it up the mucous membrane was seen to be intact; about 12 inches from the anus was a small spot of capillary injection.

Arteries.—The aorta, iliac and femoral arteries were removed and carefully examined; there was no sign of laceration or bruising of either their coats or of their cellular sheaths. The cut ends of the right femoral and profunda arteries were closed by firmly adherent clots. The internal circumflex artery was found not to have been cut.

Veins.—A clot filled the right femoral vein and extended up as high as the middle of the external iliac vein. A firm thrombus was also found in the left common iliac vein, just reaching into the mouth of the internal iliac vein, filling the external iliac, the femoral, profunda, popliteal, tibial, peroneal, and internal saphena veins, and every one of the small veins cut across in making the dissection of the larger trunks. Indeed, in dissecting the limb we saw small black points—the

cut ends of occluded veins—on every surface. The veins were
distended opposite each pair of valves. The clot was black in
colour and adherent to the vein, which was apparently unaltered.
At its upper end the clot was firmer and of a brick-red colour
on its circumference. The middle sacral vein was very large,
and entered the pelvis through the left sacro-sciatic foramen,
and, like the vena cava, contained fluid blood.

Stump.—The flaps were united over their whole surface except at posterior outer angle by a thin layer of soft lymph; no
pus; the sinus under Poupart's ligament empty. Slough had
not increased in size. Acetabulum entirely covered with granulations. Iliac glands enlarged; matting of tissues around external iliac vessels.

Remarks.—The points to be noted in this case are the following, and they may be dealt with in their order:—

    1. The nature and course of the disease.
    2. The reason for amputation.
    3. The method of amputation.
    4. The method of arresting hæmorrhage.
    5. The use and failure of the antiseptic system.
    6. The deep sutures.
    7. The thrombosis.

    1. *The nature and cause of the disease.*—Morbus coxæ in
a strumous individual is so common that I need only allude
to two special points presented in this case. The first is the
spontaneous fracture of the neck of the femur before admission,
which rendered excision of the head of the bone rather more
difficult than usual;* and the second is the fact that the acetabulum, which was bare and dry at the time of the excision,
rapidly became covered over by granulations after that operation,
although only two small fragments of bone were ever noticed
to escape with the discharge. The failure of free drainage to
secure closure of the sinuses, and the alteration in the nature
of the discharge after the excision, are also, perhaps, worthy of
note. In reference to the former I ought to add that at the
time of the amputation, the upper end of the femur was
covered over with fibrous tissue, so that there was no bone
disease to perpetuate the discharge.

* An instance of this is recorded by Mr. Pick in the 'Lancet,' 1868.

2. *The reasons for the amputation.*—The abstract of the case is so abbreviated that it possibly fails to show these. Excision of the bone and free drainage having failed to stop the discharge, a new inflammation, with great œdema and infiltration and pyrexia, having been lighted up, while hectic had gradually developed, and the urine become albuminous, there appeared no hope for the patient's life but in amputation of the limb; and looking at his age and the good condition of his digestive organs, it was felt that this operation held out some hope.

3. *The method of amputation.*—Having previously excised the head of the femur, amputation by the oval method was very readily performed. The advantages it possessed over the flap amputation are obvious and important. The inner side of the thigh was left intact, and was of great use in supporting the heavy posterior flap, and, had the man lived, would have aided in the nutrition of the part. I found post mortem that I had not severed the internal circumflex vessels, which would have been certainly cut in the ordinary operation. By a very simple calculation I found that by this means I had saved from 18 to 24 square inches of wound surface, and by so much, therefore, did I lessen the shock and the subsequent danger of the operation.* When amputation has to be resorted to after excision, or when the joint is so disorganised that disarticulation has already occurred or is easily accomplished, it will be surely worth while to practise the oval rather than the flap method. Several surgeons have suggested the oval operation, *e.g.*, Cornuau, Malgaige, Belmas, Scoutetten, and Guthrie, and the last named performed it. M. Verneuil also appears to have done so, but he commenced his incision over the femoral artery, and ligatured the main vessels before cutting the deeper structures. In the 'Lancet,' vol. i., 1866, a case of Mr. Lee's is recorded in which he made a longitudinal exploratory incision over the outer side of the hip, and then, on discovering the extent of the disease, he converted it into an oval amputation by carrying the knife round the thigh at the lower end of the longitudinal incision. But this case does not appear to have attracted the attention it deserves, and I have not been able to find any other case in

---

* Another advantage of this method is the greater ease with which antiseptic dressings can be applied to the stump.

which this method was employed by British surgeons. Mr. Furneaux Jordan has recently published a case ('Lancet,' vol. i., 1879) in which he first enucleated the femur, and then removed the limb by the oval method; and he has favoured me with a note in which he tells me that he operated in the same way on a case some years ago.

4. *The method of arresting hæmorrhage.*—The real interest of this case centres in the use of Mr. Richard Davy's lever. This instrument is described and its use figured in the 'British Medical Journal,' 1878, vol. i. It was first employed by Mr. Davy, at the Westminster Hospital, in 1877, and next in the case which is the subject of this paper, which was the first time it had been used on an adult.* The lever is a straight cylinder of hard wood, about 2 feet long, most carefully turned and smoothed. Its circumference is 2 inches, but at each end it is enlarged for a distance of 3 or 4 inches to a circumference of $2\frac{3}{4}$ inches. Previous to its use the rectum must be emptied, and should be free from stricture and ulceration. The lever, having been carefully lubricated, is passed up the rectum towards the sacral promontory, and then the handle is inclined to the sound side and gently raised until the pulsation in the femoral artery ceases; it is then steadied by being grasped along the inside of the sound thigh. In my case this was most easily accomplished, and the command of the vessels complete. During the incision, only the blood lying in the vessels severed was lost; afterwards a small quantity was allowed to escape, to enable us to recognise the smaller arteries; but when all the blood in the tray was measured it amounted to less than 3 oz., and in this amount there was of necessity some sawdust and serum. There was no evidence during life or after death that the lever caused any injury of the rectum, peritoneum, or vessels. The lever has since been used by Mr. H. Marsh; by Mr. Davy in a case of iliac aneurism, where pressure was kept up for 20 minutes; and also in a case of amputation of the hip by Mr. Armstrong of Greenwich, and again by Mr. Cadge of Norwich, in a similar case. Mr. Cadge writes:—'The lever answers admirably, and I was surprised to see how easily and perfectly and with how little pressure the common iliac was

---

* The lever is made by Messrs. Wright & Co., New Bond Street.

controlled.' In all six of these cases its use has been easy,
efficient, and without ill result.* The objections that may be
urged against it are: (1) that it is a coarse proceeding; (2) that
it may stretch or tear the rectum or meso-rectum; (3) that in
all cases the common iliac vein will be compressed as well as
the artery. To this it may be replied that in the six recorded
cases its use has been very easy indeed, and without injurious
effect, and that during life both the rectum and meso-rectum
are very elastic; there is no need to fear the effect of pressure
on the vein—veins are constantly exposed to firm and continuous pressure in the treatment of aneurisms by instrumental
compression, and without injury. It must, however, be admitted that in disease of the rectum (assuredly a most rare
condition where such a formidable operation would be performed) its use would be inadmissible or impossible, and in
certain cases of very short meso-rectum it might be impracticable to compress by its means the *right* common iliac. The
employment of great gentleness and care will guard against
mischief being done. The question now arises whether this
lever possesses advantages over the more common plans of
treatment—digital compression of the common iliac artery,
and compression of the aorta by Lister's (or Pancoast's) tourniquet? Digital compression is only possible in children, and
in adults with very lax belly-walls, and is then difficult to
maintain, especially if the operation is at all prolonged.
Aortic compression is undoubtedly as a rule safe and efficient,
but it is well to remember that Mr. Spence has recorded a case
('Edinburgh Medical Journal,' 1864) in which he failed to control the artery by means of Lister's tourniquet; and Mr. Bryant
mentions ('Medical Chirurgical Transactions,' vol. lv.) that in
employing the tourniquet in the treatment of a case of aneurism
he was once three hours before he was able to compress the
artery efficiently; and the result of twelve hours' compression
was acute peritonitis, ecchymosis of parietal peritoneum, mesocolon, and mesentery, while a coil of jejunum was so nipped

---

* Mr. R. Davy informs me that up to this date (July 29, 1879) the lever
has been used on three other occasions, in amputation at the hip-joint, making
in all nine cases, in eight of which the right artery has been compressed. Its
employment has been attended with such success that the total amount of
blood lost in the eight amputations is only 16 oz. No ill result has followed
from the use of the lever in any instance.

H

that intestinal obstruction was caused, and the tissues around the aorta are described as loaded with effused blood. The advantages that can be claimed for the lever are the following:—

a. The disturbance of the circulation is less; the flow of blood through the sound limb and half of the pelvis is not interfered with.*

b. The lever does *not interfere with the abdominal respiratory movements,* nor is its use interfered with by them, as any instrument acting through the abdominal walls is. It can also be used in cases of *rigid abdominal walls, great obesity,* and in many cases of *abdominal tumour,* where the tourniquet could not be employed.

c. Less of the abdominal contents are compressed. The parietal peritoneum, omentum, and mesentery escape entirely, while it is improbable that the small intestine would be implicated; only one side of the rectum with its peritoneal covering is compressed.

d. Less pressure is required to control the stream of blood through the common iliac than through the aorta.

e. The common iliac artery is compressed more securely by the rounded lever in the hollow between the psoas and the lumbo-sacral promontory than is the aorta by the curved tourniquet pad against the convex spine.

f. It is more rapidly and easily manipulated. The slight movements of raising and depressing the hand are more easily and quickly accomplished than is the working of a screw.

g. The lever is far *cheaper* and *more durable* than the tourniquet, and if not at hand in any emergency its place can easily be supplied.

The first four of these advantages are obvious and real, and appear sufficient to demand a careful and patient trial of the lever; the last three advantages, although making the demand more urgent, would not of themselves, perhaps, overcome the surgeon's rational objection to 'acting in the dark.'

5. The use and failure of the antiseptic system.—It has been stated (Stimson's 'Operative Surgery') that it is impossible to use Lister's antiseptic dressings after this amputation. I found, however, no special difficulty. I used a shield-shaped

---

* The tourniquet must often, if not always, stop the flow of blood through the inferior mesenteric artery, and still further disturb the circulation.

dressing, with the base over the buttock, and the other end brought over the stump to the groin. This was easily retained in place by a many-tailed bandage; one broad bandage was fastened round the waist, and to the back of this were sewn several double strips of gauze bandage; these 'tails' were brought round the stump and fastened to the band in front. The failure in my case was, I believe, due to not making the wound aseptic at the operation. I had intended to scrape out all the old sinuses, but, owing to the hurry necessary at the end of the operation, I could only well syringe them with carbolic lotion. Although possible after amputation by flaps, it is far easier to keep the wound aseptic when the limb is removed by the oval method.

6. *The deep sutures.*—In the attempt to get union by first intention there is nothing more important than to secure perfect apposition and immobility of the parts. Evidently the common superficial sutures fail in this respect in the case of the flaps in amputation of the hip, and it appeared to me that it would be well to have the flaps supported more firmly than bead or quill sutures do, and by something which could be *rapidly* applied. I therefore asked Messrs. Wright to make for me some suture pins, which answered these purposes very well. Each pin is of soft steel, about the size of a knitting-needle, 9 inches long, and brought to a flat spear-point at one end, while the other is fixed to the centre of an oval piece of vulcanite $1\frac{1}{2}$ in. × 1 in. and $\frac{1}{8}$th in. thick, a second oval pad of pure black rubber of the same shape and size being supplied with each pin. It is used thus: the pin is thrust through the flaps until the vulcanite pad is in contact with the skin, the rubber is then slipped down over the pin to the other flap with a sufficient degree of firmness, and the pin divided half an inch above it by wire nippers. The elastic rubber grips the pin, and the flaps are securely held between the oval pads. I employed three of these sutures, introduced about 2 inches from the edge of the flaps, avoiding the great vessels. The sutures can, of course, be withdrawn with the utmost ease by slipping off the rubber, and then seizing and drawing upon the vulcanite. The pins are readily made aseptic. They succeeded very well in this case, preventing all quivering, retaining the flaps in accurate apposition, and preventing retention of exudation. Indeed, they acted so well

that, while being dressed, the man was able to turn over on to his sound side and raise his stump without any assistance—the stump moving as one mass.

In comparing these sutures with different forms of wire suture it is to be noted that they secure fixity of the flaps without tightness. Wire sutures only hold parts firmly and immovable when applied *tight,* and the nipping of the parts is never good, and may be injurious; the inflexible pins, however, fix them steadily and securely, though the rubber pad is only gently pressing upon the skin.*

7. *The thrombosis.*—The cause of the man's death was, no doubt, the grave blood change evidenced by the widespread thrombosis. It is to be noticed that this condition was very nearly symmetrical, and that it was not limited to a part or the whole of one vein, but extended into every small radicle of vein in the left leg. That the clotting began in the common iliac vein is indicated by the brick-red colour of the clot there, while that in the smaller veins was black; and also by the distension of the veins at each pair of valves. The absence of œdema, however, is opposed to this view, and is a striking fact in the case. It cannot be held that the slough in one corner of the stump was due to thrombosis, for it was very limited in extent, and appeared before there was any evidence of the plugging of the veins; probably it was occasioned by bruising of the infiltrated and unhealthy tissue. Although unable to speak positively as to the cause of the thrombosis, two possible factors are evident: (1) the condition of debility and prostration produced by the disease and the operation; and (2) the absorption of carbolic acid, as shown by the discoloured urine.

In favour of the former view is the well-recognised occurrence of the event, but against it are the facts that the man's general condition underwent a sudden change—he did not gradually become weaker after the operation; and then the wide extent of the thrombus. In favour of its being due to absorption of carbolic acid are the facts that the fatal symptoms set in coincidently with the excretion of ' carbolic acid urine; '

* Since this paper was read these pins have been used with good effect in two cases, under the care of my colleagues Mr. Cowell and Mr. Davy. In Mr. Cowell's case the pin was found to be very useful in restraining intermediary hæmorrhage which set in, for by depressing the rubber pad the flaps could be firmly pressed together.

that carbolic acid, when acting directly upon the blood, is known to produce thrombosis, and that in some of the fatal cases of carbolic acid poisoning the blood has been found firmly coagulated in the heart and large vessels.* Against this view it may be urged that in very many cases of carbolic acid poisoning the blood has been dark and fluid post mortem—this may be accounted for by the asphyxial symptoms usually attending these cases; and also that this effect has not hitherto been noted, although cases of absorption of carbolic acid from wounds are of frequent occurrence.

It is certain that the thrombosis was in no way connected with the use of Davy's lever, for the clotting did not extend up to the part of the vein which was compressed, and no change whatever could be detected in the vessel at that spot.

AMPUTATION OF THE RIGHT LOWER LIMB AFTER EXCISION OF THE HIP—USE OF DAVY'S LEVER.

By G. W. ARMSTRONG, M.R.C.S. Eng.,

Surgeon to the Royal Hospital School, Greenwich.

RICHARD C., aged 11 years, a boy in this school, who had always been previously in good health, although of delicate appearance, came to the Infirmary on February 2, 1878, complaining of pain in the left ankle. There was some swelling about the joint, but no redness, nor much pain on moving it. He said he thought he had sprained it, but did not remember doing so. There was a slight graze on the right knee, which was healing and looked healthy. Fomentations were applied to the ankle.

On the next day he had sharp feverish symptoms. His temperature was 104·4°; tongue white and furred. The case was judged to be one of acute rheumatism, and ten grains of salicine were ordered to be given every hour. On February 4 the morning temperature was 101·3°, falling in the evening to 101°. The ankle was very red, and acutely painful. On the

---

* Ogston and Zimm, quoted by Bœhm, Ziemssen's ' Cyclop.,' vol. xvii., p. 531.

5th the left middle finger was very red and swollen; temperature at night 103°.   On the 7th, he complained of pain in the right hip, thigh, and leg.   On the 9th, the heart's action was tumultuous and irregular, and pain in the right ankle was complained of.   The dose of salicine was increased to fifteen grains every hour, which was continued until the 12th, when, it being evident that pus had formed in the left ankle and in the metacarpo-phalangeal joint of the left middle finger, the salicine was discontinued and free incisions were made into these joints, a considerable amount of pus being evacuated.   On February 16, he had pain in the right heel and back of the right knee; he improved slightly, sleeping better, and having less pain, until the 23rd, when the right knee became painful, but the pain passed off.   It returned on March 1, and the joint swelled and became full of fluid.   On March 3, redness and swelling were noticed below the left internal malleolus.   On the 6th, an abscess, which had formed without pain just above the left elbow-joint, was opened.   On the 10th, an abscess behind the left ankle-joint was opened.   On March 18, the right knee was freely incised on the inner side, and a good deal of pus let out.   On the 26th, he was placed on a water-bed, there being a sore on the lower part of the sacrum; and, although great care was taken, others formed over the spines of the dorsal vertebra.   On April 7, there were petechiæ on the left arm.   An abscess which had formed about the right hip-joint was opened, the incision being over the great trochanter; six ounces of pus were evacuated.   On the 11th, an opening was made on the outer side of the right thigh, three inches above the knee.   There was free discharge from all the openings in the right limb.   On April 25, a bedsore formed on the right ankle.   He was very weak and exhausted, but took nourishment and wine well.

From this date he gained somewhat in strength, and the abscesses on the left side healed; the sores on the back became cleaner, but did not heal; the discharge from the right thigh and hip was very great, and until December he continued in this state, when Mr. Johnson Smith, of the Seamen's Hospital, and Mr. Pink saw him, in consultation with me. The right knee was ankylosed; and, no bare bone being felt in the shaft of the femur, it was determined to try if excision of the hip would preserve the limb.   Accordingly, on December 17, Mr.

Johnson Smith, Mr. Pink, and Mr. Purvis kindly assisting, Dr. Williams administered chloroform, and I made an incision about six inches long over the great trochanter, and through this removed the head and both trochanters of the femur, and gouged away the whole of the acetabulum. Some strips of lint were placed in the wound, and its edges were brought together by sutures. The limb was laid between sandbags, and a bag, containing six pounds of shot, was attached to the leg by strips of plaster, and hung over the end of the bed, to extend the limb. He slept well after the operation, under the influence of two doses of fifteen minims each of tincture of opium.

December 20.—He was put under the influence of chloroform, and the sheets were changed and the back dressed, &c. There was very great discharge of pus from the wound at the hip, and also from the inner side of the thigh. This method of dressing, under chloroform, was repeated every two days until January 21, 1879, on which day he was so weak that it was not deemed safe to administer chloroform. Mr. Johnson Smith and Mr. Pink saw him with me on this day; and we determined that, if he were stronger on the morrow, we would amputate the limb.

On January 22, kindly assisted by Mr. R. Davy, of the Westminster Hospital, Mr. Johnson Smith, Mr. Pink, and Mr. Bampton, the boy was anæsthetised by Dr. Williams. Mr. Davy controlled the hæmorrhage by his peculiar method. He introduced a wooden staff, about two feet in length and three-fourths of an inch in diameter, having a bulbous extremity, into the rectum for a distance of nine inches; and then, using the anus as a fulcrum, compressed the common iliac artery against the pelvis. I removed the limb by antero-posterior flaps, commencing in the wound made for the excision. Fifteen ligatures were applied; and so completely did Mr. Davy's instrument prevent hæmorrhage, in both anterior and posterior flaps, that we were obliged to ask him to relax his pressure in order to discover the bleeding points. The total quantity of blood lost was measured, and was only three ounces, this being mixed with pus. The acetabulum and pelvis were examined, but no bare bone could be felt. The flaps were brought together with silk sutures.

The patient was greatly exhausted after the operation, and

was almost pulseless. Brandy being freely administered, he gradually rallied, and, after a dose of fifteen minims of tincture of opium, he slept for a time.

On the 23rd, he had passed a fair night, taking nourishment freely. There was a considerable discharge of serum from the outer part of the wound. On the 24th, there was much serous discharge. He was carefully lifted; the stump and back were sponged, and simple ointment on lint applied to both, and the sheets, &c., were changed. On January 25, the purulent discharge was offensive and copious; the scrotum and left thigh were œdematous. He was dressed as yesterday. On the 26th, the stump looked rather better; tongue moist. At his own wish he tried to eat part of a mutton-chop, which caused vomiting and slight oozing of blood from the wound. The vomiting exhausted him very much, and at night he was very weak. The voice was hoarse, and he could not clear the larynx by coughing. On the 27th, his bowels were relaxed, to check which small doses of tincture of opium were administered, and a teaspoonful of Brand's essence of beef was given every hour. On the 28th, he took the essence well; the purging was stopped. He slept fairly. There was a free discharge of pus from the stump. On the 29th, he was very weak. The side was excoriated by the discharge which ran under him. He was ordered to have as much port wine as he could take. On the 30th, the œdema of the left limb and scrotum had increased, and was extending over the abdomen. He passed very little urine. On the 31st, the pulse was hardly perceptible. There was a glazed swelling over the left groin, with slight blush. He sank gradually, and died at 5 P.M.

February 2.—I was allowed to open the abdomen; but, the father wishing to be present, I could only make a very partial examination. The peritoneum lining the cavity of the pelvis appeared healthy, and no pus or slough appeared on incising this membrane in several places. I examined the course of the right common iliac for any sign of pressure exercised whilst controlling the hæmorrhage at the operation, but could see none; nor was any cause discovered by me, in the necessarily imperfect examination, for the œdema of the left limb. The shaft of the removed femur was atrophied, but in other respects healthy and covered by periosteum, save over a small portion, about two inches in extent, of the posterior surface, near the upper end truncated extremity of the bone.

CONTROL OF HÆMORRHAGE IN AMPUTATION AT THE HIP-JOINT.

THE following notice of the Lever appeared in the 'Lancet' on December 21, 1878 :—

Mr. Alfred Pearce Gould performed the operation of amputation at the hip-joint on the 7th inst., at Westminster Hospital. The patient was a young man, aged 28, in whom Mr. Gould had previously resected the joint. The hæmorrhage was controlled by an original device of Mr. R. Davy's, and so completely that only about three ounces of blood were lost. Mr. Davy compresses the common iliac artery by introducing a straight wooden rod, with a bulbous end, carefully into the rectum for about nine inches. The whole length of the rod is about twenty-two inches. It requires, of course, considerable knowledge to apply this instrument accurately and to use it harmlessly. But, in skilful hands, the slightest elevation or depression of the handle, when once the instrument was brought to bear on the vessel, was enough to stop or to allow the flow of blood.

We were struck with the complete anæmia of the stump when Mr. Davy lightly raised the handle of the stick. Notwithstanding the slight amount of blood lost the patient, unfortunately, died on the fourth day after the operation. The post mortem examination showed that the parts where pressure had been applied to control hæmorrhage were quite uninjured. The chief morbid appearances were extensive thrombosis of the veins of the opposite limb, extending into the common iliac vein.

AMPUTATION AT THE HIP-JOINT.

ADELAIDE HOSPITAL, DUBLIN.

THE subjoined communication was read by Mr. B. Wills Richardson, F.R.C.S., before the Surgical Society of Ireland, and is reported in the 'Medical Press and Circular,' of April 9, 1879.

Amputation at the left hip-joint was performed on Saturday last, the 1st inst., in the Adelaide Hospital, Dublin, by Mr. B. Wills Richardson. Compression of the common iliac artery was made with Mr. Richard Davy's wooden lever passed up the rectum, which so effectually controlled the vessel that the operation was almost bloodless.

Being against rule to give the particulars of operations in recent specimen exhibitions, I must crave your permission, Mr. Vice-President, to allow me to deviate from our custom on the present occasion, and to allude to one matter only, it being not altogether devoid of novelty, and cannot fail to be of great interest to those gentlemen present, who might at any moment be called upon to amputate at the hip. The full particulars of the case I purpose publishing at some future period, but this is well worthy of immediate corroborative mention. I refer, Sir, to a method for controlling the vessels in amputation at the hip which is thoroughly effective and easy of application, the compression of soft structures being reduced to a minimum.

In the 'British Medical Journal' for May 18, 1878, Mr. Richard Davy, of the Westminster Hospital, London, described a wooden lever for compressing the common iliac artery in amputation at the hip. It appeared to me so simple and applicable that I resolved to use it in preference to the abdominal compressor attributed to Mr. Lister, and can now confidently recommend it as an instrument capable of rendering amputation at the hip one of the most bloodless capital operations in surgery, with a minimum amount of risk to the abdominal or pelvic contents. The lever (exhibits it) is turned out of a piece of boxwood, its length being about $22\frac{1}{2}$ inches. Both ends enlarge, each enlargement being $3\frac{1}{2}$ inches long, the diameter of each being 1 inch. They are oval at the

free end, and at the other diminish in calibre, and become continuous with the shaft, which has a uniform diameter of $\frac{3}{4}$ of
an inch.

I here take the opportunity of saying a few words in grateful acknowledgment of the efficacious manner in which the
lever was handled in this case by Mr. William Thomson, for my
colleague Dr. Barton, who was at the time recovering from
lumbago, and incapable of the stooping posture essential for its
proper working.

The rectum being empty, two ounces of linseed-oil were
thrown into it. One end of the lever having been oiled also,
Mr. Thomson carefully, with a rotatory motion, passed it through
the anus and up the rectum, until it was felt above the brim
of the true pelvis by pressing the abdominal wall. An Esmarch's bandage, which had been previously applied to the
whole limb and around the pelvis, to save for the child as much
as possible of the blood that had circulated in the limb was
removed. Mr. Thomson, at the commencement of its unrolling,
brought forward the protruding end of the lever, making a fulcrum of the perineum, thereby throwing the other end backward
and compressing the common iliac artery in the groove between
the bodies of the lumbar vertebræ and the psoas magnus muscle.
Pulsation at once ceased in the femoral artery, and the operation was then commenced and soon completed. So effectual
was the control of the iliac artery that at the very utmost not
more than two ounces of blood were lost. Indeed, it is probable
that when the operation was finished the child, thanks to the
elastic bandage also, had more blood than she possessed before
it. It is a source of great satisfaction to me to be able to report that, with the exception of some vomiting the day after
the operation, seemingly caused by absorption of carbolic acid,
the urine being then dark olive-brown, nearly black, from the
presence of the drug (exhibits specimen), she has not had a
symptom to excite alarm for her safety. The urine was tested
by our Assistant Physician, Dr. Walter Smith, who reported to
me that, although free from blood and albumen, there was no
doubt of the presence of carbolic acid. Those who believe in
the deleterious influence of bacteria on the system after operations might argue with much plausibility that the circulation

for a few hours of carbolic acid in the blood may have had some influence in leading to her present favourable condition.

I should observe that the lever used was the smaller (in diameter) of the two exhibited; the other, the one described above, being a little too large for the child. It was turned for me by Mr. Prescott, of South King Street, Dublin, and has not warped, which the larger has done.

Before I sit down I had better state that Mr. Davy mentions, in a letter I received from him dated the 20th of February, that he was using the lever in compressing the common iliac artery for the cure of an external iliac aneurism.

AMPUTATION AT THE HIP-JOINT—USE OF LEVER TO COMPRESS COMMON ILIAC ARTERY.

MANCHESTER ROYAL INFIRMARY.

(Under the care of Mr. WALTER WHITEHEAD, *Surgeon to the Infirmary.)*

J. T. S., aged 26, suffered from morbus coxarius since he was four years of age.   On September 22, 1879, the head and several inches of the shaft of the femur were excised.   On October 11, secondary hæmorrhage occurred, and recurred on the 12th and 15th. The bleeding was so profuse, and proceeded from situations so difficult of access, that, taking into consideration the uselessness and the general unsatisfactory prospect of the limb, it was decided, in consultation with Mr. Bradley, to amputate. Immediate action being imperative, and Lister's abdominal tourniquet not at the time available, it was suggested to try the lever.   Mr. Bradley, acquiescing, consented to superintend its use, and selected for that purpose the roller of an ordinary small window-blind.   This was made smooth and somewhat rounded at one end ; and, after being well smeared with oil, was passed up the rectum and pressed on the pelvic wall until Mr. Bradley was satisfied, by the absence of pulsation at the groin, that the common iliac as under perfect control. The amputation was leisurely conducted, and the best flaps made that the sinus-riddled tissue would permit. The operation was performed without the loss of any blood beyond what might reasonably be considered to come from the tissues below the point of compression, and not in all amounting to more than an ounce. The limb had, previously to amputation, been subjected to the use of Esmarch's bandage. The man died on the table, without there being, however, any very obvious cause for immediate death.

The special object for reporting this case is to acknowledge the satisfactory and complete aid that was afforded by the use of the lever for compressing the common iliac during the operation—a procedure suggested by Mr. Richard Davy in 1874 ('British Medical Journal,' May 18, 1878, p. 704), and made use of by him in 1876.

One great advantage of the lever is, that it can be extemporised out of so many ordinary and available articles of domestic furniture ; an ordinary walking-stick, or even the rail of a bedroom chair, would equally serve the purpose.

It can be imagined that in very stout people Lister's tourniquet might fail to accomplish its object ; but, unless there should be a tight stricture of the rectum, or, as suggested by Mr. Bradley, a short mesocolon, preventing much lateral deflection of the rectum, it is difficult to apprehend any other condition or circumstance which would render the lever inapplicable.

MR. WILLIAM STOKES' PAPER,

AND DISCUSSION THEREON, AT THE CLINICAL SOCIETY OF LONDON, *Friday, April* 23, 1880.

AMPUTATION THROUGH THE HIP-JOINT.—Mr. William Stokes (Dublin) read notes of this case. Advantage was taken of the facilities afforded for controlling hæmorrhage by means of Davy's rectal lever, and the best results were thereby obtained. B. Pemberton, aged 42, tall, muscular, and well-nourished, who had been successively a shoemaker, sexton, mason, and cabman, was admitted to the Richmond Surgical Hospital, Dublin,

August 29, 1879.   He then exhibited the signs of advanced
coxo-femoral arthritis.   He stated that, in 1846, being then
nine years old, he fell on the left hip.   For long after the injury
affected him, and he was attended in hospital by the late Dr.
Hutton, and subsequently came again under treatment for three
months.   At the end of this time, but for a slight halt, he was
apparently quite well.   In 1852, the joint was again injured,
and troubled him during the next five years.   In 1878, he was
thrown from a car, being then a cab-driver, and was at once
taken to the hospital in consequence.   After seven weeks he
went out, greatly relieved, but was forced to resume his bed in
April last year, and left it only to be conveyed to the hospital.
On admission he was deplorably ill.   The least motion of the
joint caused exquisite pain; even the weight of the bed-clothes
was unendurable.   The face wore an anxious expression, and
there was extreme exhaustion from pain and sleeplessness.
Temperature varied between 99° and 102° Fahr.   The limb
was much everted; shortening apparently great, but really
only three-quarters of an inch.   Three sinuses existed along
the outer side of the limb, two pointing up and in toward
the acetabulum, one downward.   Probing the former revealed
denuded and softened bone.   This could not be felt in the third
sinus.   From all came a copious, almost constant, thin, sanious,
watery, purulent discharge; a large bedsore existed over the
sacrum.   It was necessary to remove the urine twice a day
with a catheter.   Defæcation caused much distress.   Food was
not relished nor desired.   There were profuse night-sweats.
Altogether his condition called for prompt operative interference
if life were to be preserved.   Mr. Stokes naturally considered
the relative merits of resection and amputation in this case,
and decided against the former, in consequence of the great
chronicity of the case, the probable (in truth, almost certain)
existence of osseous disease below where the section should be
made in excision, the age and utterly exhausted condition, from
discharge and pain, of the patient.   He accordingly determined
in favour of amputation at the hip-joint, and the result fully
justified the course adopted.   He further determined on adopting
Davy's method of preventing hæmorrhage in the performance of
this particular amputation during the operation.   On October 1,
he performed the operation.   The limb was first elevated for

some minutes, an Esmarch's bandage applied, then the patient
was etherised, and, lastly, the lever introduced by Dr. Thomson
without difficulty. It was at once interesting and pleasing to
observe that, on the point of the instrument being brought
over the common iliac artery, the slightest elevation of the
handle caused a complete cessation of arterial pulsation in the
condemned limb. It is not too much to say that the amputation was by this means rendered almost a bloodless one. The
operation was done by the formation of an antero-external
flap, made by transfixion, and a postero-internal one, made by
incision from without inwards. This part of the procedure did
not occupy more than a few seconds. The femoral artery was
secured by a carbolised silk ligature, all others by catgut.
The strictest antiseptic precautions by Lister's method were
observed both during and subsequent to the operation. Immediately after the operation the temperature of the patient was
98·8° Fahr., pulse 136, and respiration 32. The pulse being
very weak, some brandy, with fifteen minims of ether, was
given, and the pulse became gradually better, though still
remaining at 136. Morphia was then given, after which the
patient slept for some hours. At 10 P.M. the temperature was
99° Fahr., pulse 130, and respiration 32. The patient did
not complain of pain; there was no sign of hæmorrhage or
other trouble. October 2.—He passed a quiet night until
5 A.M., when a severe rigor occurred; a second one occurred fifteen minutes subsequently; but these did not recur. 9. A.M.—
Temperature 97·6° Fahr., pulse 120, respiration 24. 9 P.M.—
The patient slept well for about four hours during the day, and
took a cup of chicken-jelly, also some milk and lime water.
October 3.—He passed a good night, having slept during the
greater part of it; temperature 99° Fahr., pulse 112, respiration 24. The wound was dressed; it looked healthy, and was
quite aseptic; the patient during the day took small quantities
of beef-tea, chicken-jelly, and iced milk. From this date the
progress to recovery was uninterrupted. The wound remained
aseptic for eight days, after which a small amount of suppuration
occurred. The femoral ligature separated on the thirteenth day.
The following note was taken on the twentieth day after the
operation :—'Wound dressed to-day; hardly a trace of suppuration. Bedsore healing rapidly. General state of the patient

very satisfactory.   Is free from all pain.   His appetite is good, and he sleeps well.'   The patient remained in hospital until the middle of December, when he returned home, the wound having completely healed.

Mr. HOWARD MARSH said that Mr. Davy's method had been adopted by several surgeons, but was not yet in general sufficiently appreciated.   He was fully alive to its value, because, shortly before its introduction, he had seen a patient die from hæmorrhage.   The femoral artery had been well secured ; but Lister's tourniquet could not be applied to the abdominal aorta, and bleeding from the posterior flap occurred.   In another case, occurring soon after, Davy's lever had completely restrained the hæmorrhage.   He had heard of no bad results following the use of the instrument, but it should be employed with care.

Mr. HUTCHINSON thought that many severe cases of hip-disease were lost from surgeons' unwillingness to amputate. He had, in four such cases, amputated, in two of them with complete success.   The patient's life in one case was certainly saved by amputation.   In all these cases, which occurred before the introduction of Davy's lever, he had tied an elastic band tightly around the abdomen, and had washed out the wound with spirit.   Listerism had since been introduced, and he should adopt it in any future case, but these four cases had not suffered from the want of it.   He had advised amputation in one case, where unwillingness to resort to it had incurred delay, through which the patient had died.

Mr. MACCORMAC thought that, in some severe cases of hip-disease and of disease of the femur, amputation was the only procedure which gave hopes of recovery.   As to the occurrence of shock after large operations, Langenbeck, speaking chiefly from his experience on the battlefield, had attributed its occurrence chiefly to loss of blood; but, in a case of his own, of amputation at the hip, the patient (a boy) had intense shock, although he had lost scarcely any blood.

Mr. BRYANT said that resection in children was comparatively successful, but in adults was frequently fatal.   He consequently thought that surgeons should more frequently advise amputation in the advanced hip-disease of adults.   He had employed Davy's lever, Mr. Davy himself applying it, with the loss of only two or three ounces of blood, in a case of amputation for tumour of the neck of the femur. As loss of blood was the chief danger to be apprehended, he usually cut down directly upon the femoral artery, cut and twisted both ends at the cut, and then applied a carbolised catgut ligature to the femoral vein.

Dr. GILBART SMITH asked whether the shock produced in Mr. MacCormac's case might not have been due to the pressure on the aorta affecting also the solar plexus.

Mr. MacCORMAC said that the pressure had been slight, and that it had been applied just above the umbilicus. He could not think it was sufficient to affect injuriously the solar plexus.

Mr. PARKER had amputated at the hip in a child thirteen years old, and had found the femoral vein plugged. He could control the arterial hæmorrhage, but could not arrest that from the veins. To it he had attributed the fatal result. Mr. Stokes' case illustrated a remark he (Mr. Parker) had made at a former discussion on hip-joint disease, that sequestra were often found loose in the joint. Had they been removed at an earlier date, possibly the patient might have recovered and his limb been saved.

Mr. STOKES thought the infrequency of the operation of amputation was due to surgeons' fear for the result; but Mr. Davy's lever had lessened the great risk of a fatal result from loss of blood. The mortality had previously been 50 or 60 per cent. The disadvantages connected with the use of Lister's tourniquet, or of Esmarch's bandage, applied around the abdomen over the aorta, were chiefly due to their interference with respiration, which Davy's lever avoided. He thought loss of blood was always present in shock, but did not know that Langenbeck considered it to be the *only* cause. Bleeding was chiefly from the posterior flap in amputations at the hip. If the case had come sooner under his notice he might have attempted to remove the sequestra from the joint. He did not think the course of the case was entirely due to their presence. The disease was very extensive, implicating not only the head of the femur but also the acetabulum.

AMPUTATION AT THE HIP—USE OF DAVY'S LEVER.

By R. Clement Lucas, B.S.Lond., F.R.C.S.,

Assistant-Surgeon to Guy's Hospital, and to the Evelina Hospital for Sick Children.

('*British Medical Journal,*' *December* 20, 1879.)

As it is by the accumulated experience of many surgeons, rather than by the extended practice of one, that an instrument is accepted or condemned, I think it right to state the result of my trial of Mr. Davy's lever for the compression of the common iliac artery during amputation at the hip-joint. The patient was a boy, aged 10 years, who had suffered from hip-joint disease for six years. When he was admitted into the Evelina Hospital there were shortening of the limb, eversion, and a gluteal abscess. I excised the head of the femur on August 4, but the patient showed little recuperative power; he suffered from profuse suppuration and a high temperature; he grew gradually thinner; and it became evident, after a time, that a pelvic abscess had formed. The rapid emaciation and pallor left no doubt as to his approaching end. I therefore determined, on November 15, to amputate.

Before placing the patient on the operating table an experimental trial of the lever was made, under chloroform, on a convalescent patient. We satisfied ourselves in this way that, with very slight pressure, the femoral artery and its branches could be completely controlled.

The patient was prepared for operation as follows :—His body and extremities (except the one to be amputated) were wrapped in cotton-wool, to keep up the temperature and favour circulation in these parts; Esmarch's bandage was then tightly applied from the foot to the middle of the thigh, to squeeze all the blood out of the limb; and ether was administered in preference to chloroform, as it has a stimulating rather than depressing effect upon the heart's action. Dr. H. Davy, the house surgeon, passed the lever between eight and nine inches into the bowel, and raised the handle till pulsation could no longer be felt in the femoral artery. I then cut an antero-external flap, commencing below the pubic spine, and ending in the old wound made for excision. The flap was dissected up until the femoral artery was exposed, and this was then seized by two pairs of forceps, divided between them, and twisted. The femoral vein was treated in the same way. Another light cut exposed the profunda artery, which was secured like the femoral before division, and afterwards twisted. The soft parts were then divided to the level of the bone. At this time a little spurt came from a branch of the external circumflex, but raising the handle of the lever appeared to control it. The amputation was completed by cutting a postero-internal flap. One or two vessels bled in the posterior flap uncontrolled by the lever, and from this I concluded that the lever was pressing at or below the bifurcation of the common iliac. My opinion of the lever is that, in action, it is efficient and preferable to that of other instruments at present employed in similar cases. At the same time, seeing how important it is in these amputations to save every drop of blood, I would strongly commend the proceeding of the dissecting-up of the anterior flap, and securing the vessels with forceps before their division. The surgeon thus, whatever tourniquet may be employed, renders himself almost independent of mechanical aid, and uses it only as an additional safeguard against loss of blood. Neither of the patients on whom the lever was tried suffered from diarrhœa or any abdominal symptom subsequent to its use. During the operation the patient's pulse gave us no cause for anxiety; and since then (now a fortnight) his general condition has improved, although the discharge from the pelvic abscess (and bedsores) remains profuse.

Before closing this note I would like to insist on the extreme importance of keeping up the temperature, and of preserving every drop of blood during this amputation—details in operating which I have more fully discussed in the last volume of the 'Guy's Hospital Reports.' Having by these means, and by the free use of stimulants during the first twelve hours, tided the patient over the danger from shock, there ought to be little further cause for anxiety, since torsion has eliminated all fear from secondary hæmorrhage, and antiseptic surgery has done much to exclude the possibility of death from septicæmia.

Spottiswoode & Co., Printers, New-street Square London.

SMITH, ELDER, & CO.'S PUBLICATIONS.

SURGERY: its Principles and Practice. By TIMOTHY HOLMES, M.A.
Cantab., F.R.C.S., Surgeon to St. George's Hospital. Second Edition. With upwards of 400
Illustrations. Royal 8vo. 30s.

ANTISEPTIC SURGERY: an Address delivered at St. Thomas's
Hospital, with the subsequent Debate. To which are added a short Statement of the Theory
of the Antiseptic Method, a Description of the Materials employed in carrying it out, and some
Applications of the Method to Operations and Injuries in Different Regions of the Body, and
to Wounds received in War. By WILLIAM MAC CORMAC, M.A., F.R.C.S.E. & I., M.Ch.
(Hon. Caus.), Surgeon and Lecturer on Surgery at St. Thomas's Hospital. With Illustrations.
8vo. 15s.

A SYSTEM of SURGERY: PATHOLOGICAL, DIAGNOSTIC,
THERAPEUTIC, and OPERATIVE. By SAMUEL D. GROSS, M.D., LL.D., D.C.L. Oxon.
Fifth Edition, greatly Enlarged and thoroughly Revised, with upwards of 1,400 Illustrations.
2 vols. 8vo. £3. 10s.

The ESSENTIALS of BANDAGING: for Managing Fractures and
Dislocations; for administering Ether and Chloroform; and for using other Surgical Apparatus;
and containing a Chapter on Surgical Landmarks. By BERKELEY HILL, M.B. Lond., F.R.C.S.,
Professor of Clinical Surgery in University College, Surgeon to University College Hospital, and
Surgeon to the Lock Hospital. With 134 Illustrations. Fourth Edition, Revised and much
Enlarged. Fcp. 8vo. 5s.

A HANDBOOK of OPHTHALMIC SURGERY. By BENJAMIN
THOMPSON LOWNE, F.R.C.S., Ophthalmic Surgeon to the Great Northern Hospital. Crown
8vo. 6s.

The STUDENT'S MANUAL of VENEREAL DISEASES. Being
a concise description of those Affections and of their Treatment. By BERKELEY HILL, M.B.,
Professor of Clinical Surgery in University College, London; Surgeon to University College
and to the Lock Hospitals; and by ARTHUR COOPER, late House Surgeon to the Lock Hospital.
Second Edition. Post 8vo. 2s. 6d.

A TREATISE on the THEORY and PRACTICE of MEDICINE.
By JOHN SYER BRISTOWE, M.D. Lond., F.R.C.P., Physician to St. Thomas's Hospital, Joint
Lecturer in Medicine to the Royal College of Surgeons, formerly Examiner in Medicine to
University of London, and Lecturer on General Pathology and on Physiology at St. Thomas's
Hospital. Second Edition. 8vo. 21s.

CLINICAL MANUAL for the STUDY of MEDICAL CASES.
Edited by JAMES FINLAYSON, M.D., Physician and Lecturer on Clinical Medicine in the Glasgow
Western Infirmary, &c. With Special Chapters by Professor GAIRDNER on the Physiognomy
of Disease; Professor STEPHENSON on Disorders of the Female Organs; Dr. ALEXANDER
ROBERTSON on Insanity; Dr. SAMSON GEMMEL, on Physical Diagnosis; Dr. JOSEPH COATS on
Laryngoscopy, and also on the Method of performing Post-mortem Examinations. Crown 8vo.
with numerous Illustrations, 12s. 6d.

An INTRODUCTION to the STUDY of CLINICAL MEDICINE:
being a Guide to the Investigation of Disease, for the Use of Students. By OCTAVIUS STURGES,
M.D. Cantab., F.R.C.P., Physician to Westminster Hospital. Crown 8vo. 4s. 6d.

AUSCULTATION and PERCUSSION, together with the other
Methods of Physical Examination of the Chest. By SAMUEL GEE, M.D. With Illustrations.
New Edition. Fcp. 8vo. 6s.

London: SMITH, ELDER, & CO., 15 Waterloo Place.

SMITH, ELDER, & CO.'S PUBLICATIONS.

DEMONSTRATIONS of ANATOMY; being a Guide to the Know-
ledge of the Human Body by Dissection. By GEORGE VINER ELLIS, Emeritus Professor of
University College, London. Eighth Edition, Revised. With 248 Engravings on Wood. Small
8vo. 12s. 6d. The number of illustrations has been largely added to in this edition, and many
of the new woolcuts are reduced copies of the Plates in the Author's work, 'Illustrations of
Dissections.'

ILLUSTRATIONS of DISSECTIONS. In a Series of Original
Coloured Plates, the Size of Life, representing the Dissection of the Human Body. By G. V.
ELLIS and G. H. FORD. Imperial folio, 2 vols. half-bound in morocco, £6. 6s. May also be
had in parts, separately. Parts 1 to 28, 3s. 6d. each; Part 29, 5s.

MANUAL of PRACTICAL ANATOMY. With Outline Plates. By
J. COSSAR EWART, M.D. Edin., F.R.C.S.E., F.R.S.E., Lecturer on Anatomy, School of Medicine,
Edinburgh. Part I. The Upper Limb. Demy 8vo. 4s. 6d.

The EXAMINER in ANATOMY: A Course of Instruction on the
Method of Answering Anatomical Questions. By ARTHUR TREHERN NORTON, F.R.C.S.,
Assistant Surgeon, Surgeon in Charge of the Throat Department, Lecturer on Surgery, and
late Lecturer on Anatomy at St. Mary's Hospital, &c. Crown 8vo. 5s.

A DIRECTORY for the DISSECTION of the HUMAN BODY.
By JOHN CLELAND, M.D., F.R.S., Professor of Anatomy and Physiology in Queen's College,
Galway. Fcp. 8vo. 3s. 6d.

A MANUAL of NECROSCOPY; or, a Guide to Post-Mortem
Examinations. With Notes on the Morbid Appearances, and Suggestions of Medico-Legal
Examinations. By A. H. NEWTH, M.D. Crown 8vo. 5s.

MANUAL of PRACTICAL and APPLIED ANATOMY, including
Human Morphology. By H. A. REEVES, F.R.C.S., late Demonstrator of Anatomy at the
London and Middlesex Hospitals, Assistant Surgeon and Teacher of Practical Surgery at the
London Hospital &c. Vol. I. With numerous Illustrations. 8vo. [In the press.

QUAIN and WILSON'S ANATOMICAL PLATES. 201 Plates.
2 vols. Royal folio, half-bound in morocco, or Five Parts bound in cloth. Price coloured,
£10. 10s.; plain, £6. 6s.

ATLAS of HISTOLOGY. By E. KLEIN, M.D., F.R.S., and E. NOBLE
SMITH, L.R.C.P., M.R.C.S. A complete representation of the Microscopic Structure of Simple
and Compound Tissues of Man and the Higher Animals, in carefully executed Coloured
Engravings, with Explanatory Text of the Figures, and a concise Account of the hitherto
ascertained Facts in Histology. The Atlas of Histology will appear in Thirteen Numbers,
each Number containing about Four Quarto Medium Plates, and the corresponding Text.
Price 6s. each Part. Parts 1 to 12 are now ready.

A COURSE of PRACTICAL HISTOLOGY. By EDWARD ALBERT
SCHÄFER, Assistant Professor of Physiology, University College. With numerous Illustrations.
Crown 8vo. 10s. 6d.

COMPENDIUM of HISTOLOGY. Twenty-four Lectures. By
HEINRICH FREY, Professor. Translated from the German, by permission of the Author, by
GEORGE R. CUTTER, M.D. With 208 Illustrations. 8vo. 12s.

London: SMITH, ELDER, & CO., 15 Waterloo Place.

SMITH, ELDER, & CO.'S PUBLICATIONS

A TREATISE on the SCIENCE and PRACTICE of MIDWIFERY.
By W. S. PLAYFAIR, M.D., F.R.C.P., Physician-Accoucheur to H.I. and R.H. the Duchess of Edinburgh; Professor of Obstetric Medicine in King's College; Physician for the Diseases of Women and Children to King's College Hospital; Consulting Physician to the General Lying-in Hospital, and to the Evelina Hospital for Children; President of the Obstetrical Society of London; late Examiner in Midwifery to the University of London, and to the Royal College of Physicians. Third Edition. 2 vols. demy 8vo. with 166 Illustrations, 28s.

A MANUAL of MIDWIFERY for MIDWIVES. By FANCOURT
BARNES, M.D. Aber., M.R.C.P. Lond., Physician to the British Lying-in Hospital; Assistant-Physician to the Royal Maternity Charity of London; Physician for the Diseases of Women to the St. George's and St. James's Dispensary. Crown 8vo. with numerous Illustrations, 6s.

SKIN DISEASES: including their Definitions, Symptoms, Diagnosis,
Prognosis, Morbid Anatomy, and Treatment. A Manual for Students and Practitioners. By MALCOLM MORRIS, Joint Lecturer on Dermatology, St. Mary's Hospital Medical School; formerly Clinical Assistant, Hospital for Diseases of the Skin, Blackfriars. With Illustrations. Crown 8vo. 7s. 6d.

ESSENTIALS of the PRINCIPLES and PRACTICE of MEDI-
CINE. A Handbook for Students and Practitioners. By HENRY HARTSHORNE, A.M., M.D. New Edition. 12s. 6d.

ELEMENTS of HUMAN PHYSIOLOGY. By Dr. L. HERMANN,
Professor of Physiology in the University of Zurich. Second Edition. Entirely recast from the Sixth German Edition, with very copious additions, and many additional Woodcuts, by ARTHUR GAMGEE, M.D., F.R.S., Brackenbury Professor of Physiology in Owen's College, Manchester, and Examiner in Physiology in the University of Edinburgh. Demy 8vo. 16s.

SPINAL DISEASE and SPINAL CURVATURE: their Treatment
by Suspension and the Use of Plaster-of-Paris Bandage. By LEWIS A. SAYRE, M.D., of New York, Professor of Orthopædic Surgery in Bellevue Hospital Medical College, New York, &c. &c. With 21 Photographs and numerous Woodcuts. Crown 8vo. 10s. 6d.

On FUNCTIONAL DERANGEMENTS of the LIVER. By C.
MURCHISON, M.D., LL.D., F.R.S., Physician and Lecturer on Medicine, St. Thomas's Hospital, and formerly on the Medical Staff of H.M.'s Bengal Army. Second Edition. Crown 8vo, 5s.

COMMENTARY on the BRITISH PHARMACOPŒIA. By
WALTER GEORGE SMITH, M.D., Fellow and Censor King and Queen's College of Physicians in Ireland; Examiner in Materia Medica, Q.U.I.; Assistant-Physician to the Adelaide Hospital. Crown 8vo. 12s. 6d.

246 Outline Drawings with adhesive backs, for Clinical Case Books.

OUTLINE DIAGRAM FORMS for CLINICAL CASE BOOKS.
For the representation of Injuries and Diseases and Physical Signs. Designed for the use of Clinical Students, Physicians, and Surgeons. By G. ROWELL, M.D., Resident Surgeon to the Leeds Infirmary. 3s. 6d.

The NOTATION CASE BOOK. Designed by HENRY VEALE, M.D.,
Assistant Professor of Military Medicine in the Army Medical School; Surgeon-Major, Army Medical Department, &c. Oblong crown 8vo. for the pocket, 5s.

NOTES of DEMONSTRATIONS of PHYSIOLOGICAL CHE-
MISTRY. By S. W. MOORE, Junior Demonstrator of Practical Physiology at St. George's Medical School, Fellow of the Chemical Society, &c. Crown 8vo. 3s. 6d.

London: SMITH, ELDER, & CO., 15 Waterloo Place.

SMITH, ELDER, & CO.'S PUBLICATIONS.

The FUNCTIONS of the BRAIN. With numerous Illustrations.
By DAVID FERRIER, M.D., F.R.S. 8vo. 15s.

The LOCALISATION of CEREBRAL DISEASES. By DAVID
FERRIER, M.D., F.R.S., Assistant-Physician to King's College Hospital; Professor of Forensic
Medicine, King's College. With numerous Illustrations. 8vo. 7s. 6d.

TABLES of MATERIA MEDICA. A Companion to the Materia
Medica Museum. By T. LAUDER BRUNTON, M.D., Sc.D., F.R.C.P., F.R.S., Assistant-Physician
and Lecturer on Materia Medica at St. Bartholomew's Hospital; Examiner in Materia Medica
in the University of London. Demy 8vo. 10s, 6d.

A GUIDE to THERAPEUTICS. By ROBERT FARQUHARSON, M.D.,
F.R.C.P., Lecturer on Materia Medica at St. Mary's Hospital Medical School. Crown 8vo. 7s. 6d.

An EPITOME of THERAPEUTICS. Being a Comprehensive
Summary of the Treatment of Disease as recommended by the leading British, American, and
Continental Physicians. By W. DOMETT STONE, M.D., F.R.C.S., Honorary Member of the
College of Physicians of Sweden, Physician to the Westminster General Dispensary; Editor of
the 'Half-yearly Abstract of the Medical Sciences.' Crown 8vo. 8s, 6d.

OCULAR THERAPEUTICS. By L. DE WECKER, Professor of
Clinical Ophthalmology, Paris. Translated and edited by LITTON FORBES, M.A., M.D.,
F.R.G.S., late Clinical Assistant, Royal London Ophthalmic Hospital. With Illustrations.
Demy 8vo. 16s.

DISEASES of the NERVOUS SYSTEM: their Prevalence and
Pathology. By JULIUS ALTHAUS, M.D., M.R.C.P. Lond., Senior Physician to the Hospital for
Epilepsy and Paralysis, Regent's Park; Fellow of the Royal Medical and Chirurgical Society,
Statistical Society, and the Medical Society of London; Member of the Clinical Society;
Corresponding Member of the Société d'Hydrologie Médicale de Paris; of the Electro-Thera-
peutical Society of New York; &c., &c. Demy 8vo. 12s.

The CAUSES and RESULTS of PULMONARY HÆMORRHAGE.
With Remarks on Treatment. By REGINALD E. THOMPSON, M.D. Cantab.; Fellow of the Royal
College of Physicians; Senior Assistant Physician and Pathologist to the Hospital for Con-
sumption, Brompton. Demy 8vo. with Illustrations, 10s. 6d.

The NATURAL HISTORY and RELATIONS of PNEUMONIA:
a Clinical Study. By OCTAVIUS STURGES, M.D., F.R.C.P., Physician to the Westminster
Hospital. Crown 8vo. 10s. 6d.

A TEXT-BOOK of ELECTRICITY in MEDICINE and SURGERY,
for the Use of Students and Practitioners. By GEORGE VIVIAN POORE, M.D. Lond., M.R.C.P.,
&c., Assistant-Physician to University College Hospital; Senior Physician to the Royal Infirmary
for Children and Women. Crown 8vo. 8s. 6d.

The CURATIVE EFFECTS of BATHS and WATERS; being a
Handbook to the Spas of Europe. By Dr. J. BRAUN. With a sketch on the Balneotherapeutic
and Climatic Treatment of Pulmonary Consumption, by Dr. L. ROHDEN. An Abridged Trans-
lation from the Third German Edition, with Notes. By HERMANN WEBER, M.D., F.R.C.P. Lond.,
Physician to the German Hospital. Demy 8vo. 18s.

London: SMITH, ELDER, & CO., 15 Waterloo Place.